National Vegetation Classification: Users' handbook

National Vegetation Classification:
Users' handbook

John S. Rodwell

Joint Nature Conservation Committee
Monkstone House
City Road
Peterborough
PE1 1JY
www.jncc.gov.uk

This reprint edition published by Pelagic Publishing 2012
www.pelagicpublishing.com

ISBN-13   978-1-907807-32-9

This book is a reprint edition of ISBN-13 978-1-86107-574-1 | ISBN-10 1-86107-574-X

Contents

List of Figures

List of Plates

Acknowledgements

This Handbook originates from the field methodology for vegetation sampling devised for the NVC project itself. My first debt therefore is to my colleagues in the research team who developed and tested that first basic protocol with its standard sample card: Professor Donald Pigott, then of Lancaster University, and the late Dr Derek Ratcliffe, Chief Scientist of the Nature Conservancy Council, Professor John Birks, Dr Andrew Malloch, Professor Michael Proctor and Dr David Shimwell and their research assistants Jacqui Paice (now Huntley), Dr Martin Wigginton, Paul Wilkins and Dr Elaine Grindey (now Radford).

As the NVC became a standard technique for the description of vegetation types and increasing numbers of staff within the NCC and outside made the approach their own, its seemed sensible to provide a simple *Field Manual* for use alongside the developing sections of the classification itself. This was very widely circulated among interested individuals and organisations and became a valuable teaching tool in the NVC training that was organised from Lancaster University. Very many people used and commented on that Manual and their suggestions have influenced the form and content of this Handbook. Among them, I am especially grateful to Dr Tim Bines, Dr Tim Blackstock, Alan Brown, Paul Corbett, Lynne Farrell, Dr Wanda Fojt, Rev. Gordon Graham, Katherine Hearn, Dr David Horsfield, Dr Keith Kirby, Jack Lavin, Simon Leach, Jane MacKintosh, Dr Jonathan Mitchley, Margaret Palmer, Dr George Peterken, David Stevens, Dr Chris Sydes, Professor Des Thompson, Derek Wells, Dr Peter Welsh, Dr Richard Weyl, Dr Bryan Wheeler and Geoffrey Wilmore.

Of critical importance within the NCC and subsequently the Joint Nature Conservation Committee was Dr John Hopkins, both in his original role as nominated officer to the NVC project and later to subsidiary contracts aimed at delivering particular NVC-related products. This Handbook is one of those and, within the JNCC team, I am particularly indebted to Susan Davies and Ali Buck for their encouragement and critical comments, and to Ed Mountford and Colin McLeod for finalising the handbook for publication, including redrawing of some figures, photograph selection and captions.

The Handbook incorporates and enlarges upon the original *Field Manual* and it does so partly in the light of experience gained on the much-expanded training programme of the Unit of Vegetation Science. The style and content benefited greatly by encouragement in the field and in discussion with a wide variety of staff, not only from the countryside and conservation agencies but also from the National Trust, the Royal Society for the Protection of Birds, MAFF/ADAS, the Forestry Commission, the Institute of Terrestrial Ecology, the Wildlife Trusts, the National Rivers Authority/ Environment Agency, public and privatised utilities, corporate industry, landscape architects and environmental consultants. People too numerous to name individually have tested various parts and prototypes of the Handbook and helped shape its final form.

In the Unit of Vegetation Science itself, I am enormously indebted to my colleagues on the training programme, Elizabeth Cooper and the Short-Courses Officer Kate Steele, and Julia Milton, who have facilitated this development of the Handbook; also to Michelle Needham, who tirelessly and cheerfully typed repeated versions of the manuscript.

Finally, there is a wider debt, because although this Handbook, like the *Field Manual* before it, has sprung out of the NVC, it is part of, and strongly dependent upon, an older tradition of European phytosociology. The original NVC contract brief stipulated that the classification should characterise plant communities roughly equivalent to Braun-Blanquet associations, and the research team took advantage of the long experience of vegetation sampling elsewhere in Europe in developing the NVC methodology. Since that time, many colleagues from the Continent have provided comments and advice on the classification and the field techniques described in the Handbook. Among them, I am especially grateful to Dr Joop Schaminée, Professor Victor Westhoff and Professor Sandro Pignatti.

We are grateful to Cambridge University Press for permission to reproduce a number of figures from the published volumes of *British Plant Communities*.

John Rodwell

Preface

With the publication for the Joint Nature Conservation Committee (JNCC) of the fifth volume of *British Plant Communities* (Rodwell 2000), a milestone in British phytosociology and our understanding of the vegetation of Britain was reached. The National Vegetation Classification (NVC) has become widely accepted as an important tool for nature conservation as well as in various other spheres.

Professor John Rodwell, who co-ordinated the NVC project and edited the five volumes of *British Plant Communities*, has prepared this handbook for JNCC based on his many years of experience. He has also drawn on the expertise of the many individuals who have been involved in the project, together with input from the expanding community of users.

As well as providing an authoritative introduction to the NVC, the handbook gives a detailed description of the NVC methods for collection and analysis of data. It also gives a brief account of some of the applications and limitations of the NVC, including guidance on NVC survey, although it is not intended as a manual for mapping vegetation.

This is one of a series of JNCC publications designed to aid and promote understanding and application of the NVC. We hope it will prove helpful and would welcome comments for future revisions.

Ian Strachan
Joint Nature Conservation Committee

1 Introduction

1.1. The purpose of the Handbook

This handbook provides a general introduction to the National Vegetation Classification (NVC). It details the methodology for sampling and describing vegetation in the field, explains how such information can be used to identify plant communities and outlines the character of the classification itself and the accounts of vegetation types it contains. It also discusses the important issues involved in carrying out an NVC survey of a site and gives a brief indication of other applications of the scheme.

The NVC was commissioned in 1975 by the Nature Conservancy Council (NCC) to provide a comprehensive and systematic catalogue and description of the plant communities of Britain. It has now been accepted as a standard, not only by the nature conservation and countryside organisations, but also by forestry, agriculture and water agencies, local authorities, non-governmental organisations, major industries and universities. It has been widely welcomed as providing a much-needed common language in which the character and value of the vegetation of this country can be understood. This handbook is intended to enlarge the community of users and broaden the application and value of the scheme.

1.2. Vegetation classification in Britain

British ecologists have generally been more interested in the structure and dynamics of vegetation than in what distinguishes plant communities from one another. Indeed, among many, there has been a deep-seated resistance to using phytosociological techniques and no consensus about how vegetation should be described or whether it ought to be classified at all. *The British Islands and their vegetation* (Tansley 1939), the only account we have of a wide range of the plant communities of the UK, has inspired generations of serious ecological study, but it was not systematic or comprehensive and self-consciously avoided a rigorous taxonomy of vegetation types.

An international excursion to Ireland in 1949 (published in Braun-Blanquet and Tüxen 1952) and a series of papers by Poore (1955) were important in encouraging a more meticulous approach to sampling stands of vegetation and were soon followed by *The plant communities of the Scottish Highlands* (McVean and Ratcliffe 1962). This study provided a systematic definition of a wide variety of vegetation types from an extensive region of Britain, related them to climatic, edaphic and biotic factors, and compared them to similar plant communities elsewhere in Europe, particularly Scandinavia.

In the years following, a new generation of research students began to use traditional phytosociology to classify the range of variation among British calcicolous grasslands (Shimwell 1968), heaths (Bridgewater 1970), rich fens (Wheeler 1975) and salt-marshes (Adam 1976), and to describe the vegetation of local areas such as Skye (Birks 1969), Cornwall (Malloch 1970), Upper Teesdale (Bradshaw and Jones 1976) and, in Ireland, the Burren (Ivimey-Cook and Proctor 1966). Workers at the Macaulay Land Use Research Institute in Aberdeen also greatly extended the survey of Scottish vegetation (Birse 1980, 1984), while visitors from elsewhere in Europe (Westhoff *et al.* 1959, Klötzli 1970, Géhu 1975, Willems 1978) provided Continental perspectives on particular British plant communities.

However, the coverage of vegetation types in such studies was very patchy, many data remained unpublished and there was still no co-ordinated overview of the range of variation in the United Kingdom as a whole. Dr Derek Ratcliffe, Chief Scientist of the NCC, drew attention to the great problems this posed for scientific nature conservation, while Professor Donald Pigott of Lancaster University was pointing out the need for a classification of plant communities for a proper understanding of vegetation ecology. From their concern, the NVC was born.

1.3. Rationale of the NVC

The NVC was intended from the start as a new classification, not an attempt to fit British plant communities into some existing scheme derived from elsewhere in Europe. The general approach adopted was phytosociological, concentrating on the rigorous recording of floristic data but trying to avoid some of the problems that can beset this method — over-scrupulous selection of samples, rejection of awkward data and a preoccupation with the hierarchical taxonomy of vegetation types.

The contract brief required the production of a classification with standardised descriptions of named and systematically-arranged plant communities. The survey was designed to be comprehensive in its coverage, including England, Scotland and Wales at the outset. More recently, since the completion of the project, NVC survey methods have been extended to Northern Ireland (e.g. Cooper *et al.* 1992). The NVC took in vegetation from nearly all natural, semi-natural and major artificial habitats, except where non-vascular plants were the dominants: only short-term leys, that is, agricultural grasslands sown for silage as part of arable rotations, were specifically excluded, although coverage of some habitats was limited (see Section 8.6).

It was also seen as vital that the NVC should gain wide support among ecologists with different attitudes to the descriptive analysis of vegetation. The classification was seen as much more than an annotated list of plant communities for making inventories and maps. It was meant to help understand how vegetation works, how particular plant communities are related to climate, soil and human impacts, what their internal dynamics are, and how they change from place to place and through time.

However, it is important to realise that the NVC does not provide the last word on the classification of the vegetation types of this country. It should be seen as a first approximation, essentially reliable but with some deficiencies in the coverage and much unexplained variation. *British Plant Communities*, the published version of the NVC (Rodwell 1991 *et seq.*), is not meant as a static edifice, but as a working tool for the description, assessment and study of vegetation. Section 8.6 also considers future development of the NVC.

1.4. History of the NVC

The NVC was funded throughout by a research contract to Lancaster University, with sub-contracts to Cambridge, Exeter and Manchester Universities with which the early stages of the work were shared. The project was supervised by a Co-ordinating Panel, jointly chaired by Donald Pigott and Derek Ratcliffe, with unpaid research supervisors from the four universities, Drs Andrew Malloch, John Birks, Michael Proctor and David Shimwell. With the appointment of Dr John Rodwell as full-time Co-ordinator, the NVC began its work officially in August 1975.

Four full-time research assistants, Dr Martin Wigginton, Jacqui Huntley, Paul Wilkins and Dr Elaine Radford, were appointed, one to each university, for the period 1975–1980. In the first phase of the work, they shared with the Co-ordinator the task of data collection, assembling over 13,000 new samples from vegetation types throughout the country in four field seasons.

The approach to data collection was simple and pragmatic, choosing representative quadrats located in stands of vegetation judged by eye to be homogeneous in floristics and structure. Quadrats of various sizes were used, according to the scale of the vegetation, and all vascular plants, bryophytes and macrolichens were recorded using the Domin scale. Sample location, altitude, slope and aspect were noted, as were the solid and drift geology and soil type, together with information on biotic influences including human impacts.

Existing samples of vegetation, from doctoral theses, scientific literature and unpublished reports, were added to the data where these were of compatible content and standard. At the close of the programme of data collection, a total of about 35,000 samples had been assembled and coded for analysis. They were distributed across over 80% of the 10x10 km grid squares in England, Scotland and Wales (see Figure 1).

A variety of multivariate techniques was used to characterise the vegetation types, notably TWINSPAN (Hill 1979), which was incorporated into the VESPAN package (Malloch 1988), designed using the experience of the project to provide flexible data analysis and display facilities. Samples were sorted only on their floristic attributes, environmental data being used later for interpretation of the groups produced. There was no rejection of nondescript or awkward samples and no tidying of tables to deliver a neater outcome. Throughout, the emphasis was on ecological meaning of the results, not a slavish adherence to statistical propriety. Periodic meetings of the team during data collection and analysis ensured that coverage of the country was as even as possible and that the definition of the vegetation types was proceeding on a consistent basis.

With the co-ordinator, the research assistants helped prepare preliminary descriptions of the vegetation types, and after their departure the research supervisors provided further material for writing the final accounts of the plant communities, work which took from 1980 to 1991. Manuscripts of sections of the work were circulated to NCC staff and other interested parties as soon as they were completed, and with the appearance of the first volume of *British Plant Communities*, the project entered its final stage.

1.5. British Plant Communities

British Plant Communities is the five-volume account of the NVC published by Cambridge University Press. In it, the classification and community descriptions are organised under major heads: *Woodlands and scrub* (Rodwell 1991a), *Mires and heaths* (Rodwell 1991b), *Grasslands and montane vegetation* (Rodwell 1992), *Aquatics, swamps and tall-herb fens* (Rodwell 1995) and *Maritime communities and vegetation of open habitats* (Rodwell 2000).

Each volume has an introduction to the project history and methods and an ecological overview of the vegetation types included. The community accounts themselves are organised in a modular fashion with: (1) the name and code number of the community; (2) its synonyms in previous accounts; (3) lists of constant species and characteristic rarities; (4) details of the floristics and physiognomy of the vegetation type and its sub-communities; (5) habitat relationships; (6) zonations and successions; (7) distribution, usually with a map on the 10 km National Grid; and (8) affinities with other vegetation types, including those described from elsewhere in Europe. For every community and sub-community there is also (9) a floristic table summarising the species frequency and abundance values characteristic of the vegetation.

In addition, each volume has keys to the vegetation types, indexes of species and community synonyms, and a bibliography. Volume 5 also contains a phytosociological conspectus of all the NVC vegetation types (see Section 1.8).

1.6. Relationships between the NVC and Phase 1 Habitat Survey

The Phase 1 Habitat Survey methodology was developed to provide a relatively rapid system for recording wildlife habitats and semi-natural vegetation over large areas of countryside (Nature Conservancy Council 1990, revised 2003). Its divisions are broad and designed to reflect the conservation interest of habitats. Even where Phase 1 habitat categories are further divided by using species codes, there is not a simple correspondence between its sub-divisions and the NVC plant communities. Each habitat type can include a number of NVC communities and, in some cases, the same NVC community may occur in several different habitat categories. Such cross-cutting is complex because the two systems are based on different approaches to the classification of vegetation.

Although the NVC is too fine-grained a scheme for broad reconnaissance surveys, it has become the standard for terrestrial Phase 2 survey and it is important to understand the relationships between the NVC and the Phase 1 categories. A comparison is given in Appendix 8 of the revised Phase 1 Handbook (Nature Conservancy Council 2003) and via the National Biodiversity Network Habitats Dictionary (http://www.nbn.org.uk/habitats).

1.7. NVC and the CORINE, Palaearctic, EUNIS and EC Habitats Directive Classifications

Several European-wide habitat classifications are or have recently been in usage. The CORINE Biotopes Classification was part of a European Union experimental programme of assembly of environmental information (Commission of the European Communities 1991). It is a catalogue of habitats and vegetation types arranged in a hierarchical scheme with the capacity for addition of new units at any level. It was derived by the accumulation of published definitions of units from the scientific literature and research reports and from information provided by Member States. The quality of definition is very variable and many definitions are not attributed to a source. It has also not been harmonised from country to country, certain habitats and vegetation types being represented more than once, according to whether they have been independently defined in different states. It is not comprehensive and its cover is uneven, both geographically and biologically. The scheme has no substantiating database. It has been expanded as the Palaearctic Habitat Classification (Devillers and Devillers-Terschuren 1996) and, more recently, partly incorporated into the EUNIS Habitat Classification (http://eunis.eea.eu.int).

Annex I of the EC Habitats Directive is a list of habitat types which Member States of the European Union are required to protect through designation of Special Areas of Conservation. This list was initially derived from an unpublished version of the CORINE Biotopes Classification produced in 1988, which differs from the published version of the CORINE Biotopes Classification. Member States have found difficulty in relating the Annex I list to the published version of the CORINE Biotopes Classification. An *Interpretation Manual of European Union Habitats* containing definitions of each of the Annex I habitat types has been prepared and published by the European Commission (European Commission DG

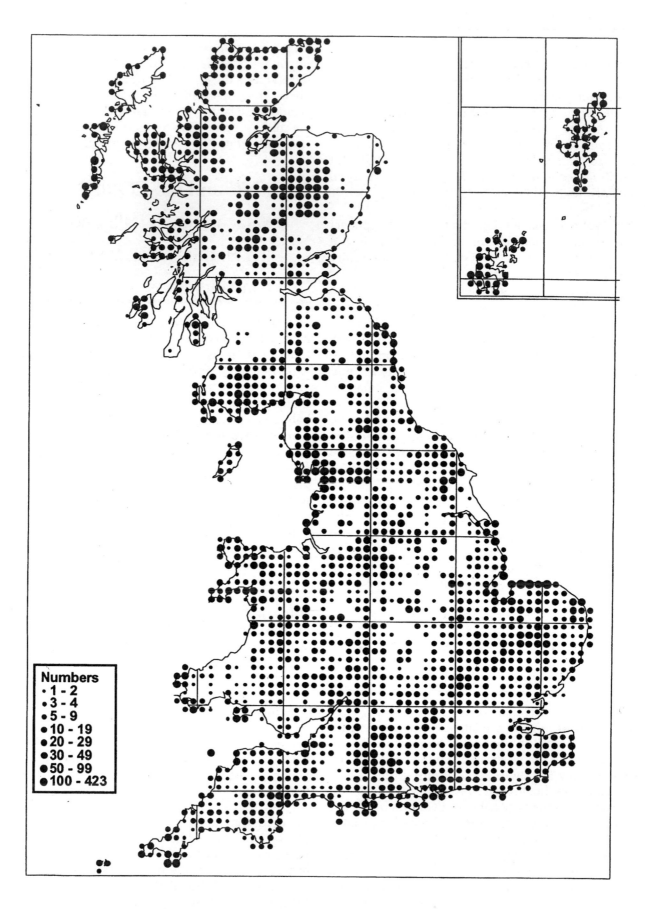

Figure 1 Geographic distribution of samples available for the NVC. Each circle in the diagram shows the number of samples in a 10x10 km grid square. The implications of this are discussed in Section 8.6. (redrawn from Rodwell *et al.* 2000).

Environment 2003) to allow experts in the EU Member States to identify individual Annex I habitats on a consistent basis. Where relevant this manual contains details of those NVC types which correspond to given Annex I habitat types. A more comprehensive review of the correspondence between the NVC and Annex I types is provided via the National Biodiversity Network Habitats Dictionary (http://www.nbn.org.uk/habitats) and in Appendix 2 of Jackson and McLeod (2000).

Such correspondences are, however, not always possible, as for example with marine habitats, which are outside the scope of the NVC project, where the boundaries of the habitat types do not correspond with boundaries of NVC communities, as for example *Tilio-Acerion* ravine forests, or the habitat types are as yet undescribed or incompletely described in the NVC, such as Mediterranean temporary ponds.

1.8. The NVC in a wider European context

Classical phytosociological data, which exists in very large quantities in many EU states and other European countries, provides a substantial basis for comparing plant communities and gaining an overview of variation among vegetation types across Europe. The standard NVC sample is essentially the same as the *relevé* (or *Aufnahme*) of the phytosociologist, and the plant communities defined by the scheme are roughly equivalent to the Braun-Blanquet Association used in phytosociological hierarchy. Also, in the descriptions in *British Plant Communities*, the affinities of each vegetation type to the most appropriate phytosociological alliance are discussed. Such comparisons are summarised in a phytosociological conspectus of all the NVC vegetation types, which is included in Volume 5 of *British Plant Communities* (Rodwell 2000) and reviewed further in Rodwell *et al.* (2000).

Meanwhile one of the benefits of publication of the NVC has been to stimulate contacts between British vegetation scientists and their European colleagues, in joint excursions, training and collaborative research. A variety of projects are now attempting to build a clearer picture of the vegetation of Europe and its vulnerability to environmental change. These are linked through a European Vegetation Survey (EVS) network that develops common survey standards and analytical software (Mucina *et al.* 1993, Rodwell *et al.* 1995), and produced an updated overview of phytosociological alliances in Europe (Rodwell *et al.* 2002). Through this network, NVC users will be able to make a substantial contribution to our understanding of the European landscape.

1.9. Use and applications of the NVC

The NVC was conceived originally as a classification scheme to help identify and understand vegetation types encountered in the field. Together with the survey methodology designed for the project, the classification is now very widely used by the UK conservation agencies and many other organisations to produce inventories and maps of plant communities on designated or threatened sites. Associated software is also extensively employed for managing databases of NVC samples and information on the distribution and extent of plant communities. Large numbers of reports of NVC surveys of individual sites have been produced and some more extensive overviews of regional or national resources based on such surveys. The NVC played an integral role in developing the *Guidelines for selection of biological SSSIs* (Nature Conservancy Council 1989) and is now a key tool in the assessment of sites in regional, national and international perspectives.

In addition to such basic applications, however, the NVC is also widely used now as a framework for scientific research into the relationships between plant communities and the environmental factors which influence their composition and distribution. Some such studies have been pursued for their intrinsic ecological interest; in other cases, the NVC has been employed to help devise programmes for managing vegetation types or individual plant species under threat. Investigations of other biota in particular habitats, such as fungi, soil bacteria and invertebrates, are also making use of the NVC as a framework for sampling, description and experimentation.

Although the NVC itself is not a monitoring tool, it is also being used to help furnish protocols for particular monitoring programmes and to develop a conceptual basis for understanding the purpose and practice of monitoring. The predictive capacity of the NVC means that it can also serve as a basis for developing management options for sites or landscapes and as a framework for restoration and design guidelines. Rodwell (1997) discusses the uses and limitations of the NVC in relation to monitoring (see Plate 1).

2 Locating samples for vegetation survey

2.1. Delimiting homogeneous stands

The first step in using the NVC in the field is to learn how to delimit stands of vegetation that are homogeneous to the eye in floristics (species composition) and in physiognomy (structure, including things like the patterned arrangement of species over the ground and vertical layering).

There is nothing mysterious about recognising homogeneity: it is not necessary to make any prior judgements about how much or how little vegetational variation is likely to be encountered in an area. Nor is there any need to look for the presence of particular indicator species to separate one stand of vegetation from another. Recognising homogeneity is a much more general visual skill: important things to look for are uniformity of colour and texture in the vegetation, repetition of any patterning over the ground and consistency of vertical layering (see Plate 1).

Nonetheless, vegetation can be very complex and field experience undoubtedly helps in differentiating stands one from another. In the first place, the scale of patterning in vegetation is very variable, from extremely fine (in, say, closely grazed pasture) to very coarse (in some woodlands), and it takes a little time to adjust the eye to an appropriate level for appreciating the variation. Then, there may be physiognomic differences within stands that are otherwise floristically uniform: the gregarious growth of some dominants, such as rank grasses in ungrazed swards, can be misleading, as can the recent occurrence of even a little grazing or mowing in part of a stand. In visiting similar areas at different times of year, phenological changes need to be taken account of too, because the switch from spring to summer dominants may markedly affect the appearance of the vegetation.

A very good general rule in examining the vegetation of a site and trying to delimit stands is not to make over-hasty judgements. It is vital to walk over a substantial part of a site first, gradually gaining a picture of the pattern and scale of variation, rather than to spend a lot of time in the first recognisable stand and then accommodate further experience inflexibly to that first impression. In many large sites, there are repeating patterns of vegetational variation such that one detailed traverse will enable a typical range of stands of different vegetation types to be delimited, although this will probably not be clear until part way through an initial general look at the whole.

Aerial photographs, especially those in full colour, can be very useful in delimiting boundaries between stands provided these are ground-truthed. In such a cloudy climate as Britain's, however, shadows and reflectance variations related to slope and aspect can be very deceptive when trying to interpret aerial photographs. Structural variation, for example, related to burning or grazing, may also be clearer than floristic differences.

2.2. Locating representative samples

Within homogeneous stands, representative samples are located through subjective choice by the surveyor (see Figure 2). Provided this selection is not influenced by a tendency to include especially rich mixtures of species or oddities of composition and structure, such subjectivity is quite acceptable. If samples have been taken from heterogeneous areas or over vaguely defined boundaries, the diversity of species composition should become apparent when the data are sorted and collated in a floristic table.

2.3. Boundaries and ecotones

Delimiting homogeneous stands is a question of recognising and avoiding boundaries between vegetation types (see Figure 2 and Plate 2). It is not necessary to know initially what the vegetation types are to be able to do this. However, boundaries are frequently rather diffuse with one vegetation type passing gradually into another, as where uniform pasturing of grasslands occurs over a sequence of soil types or where there is diffuse flushing in a sward. Then it may be very difficult to decide where one vegetation type ends and

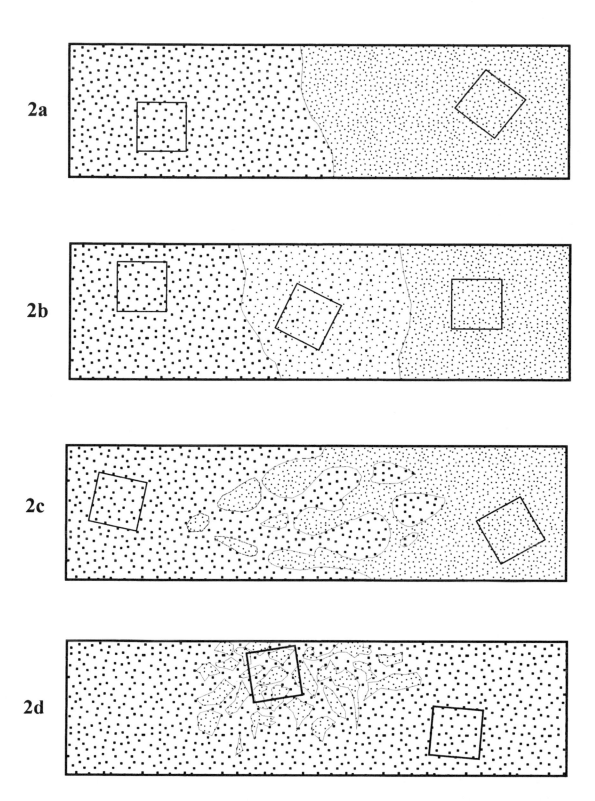

Figure 2 Delimiting homogeneous vegetation stands when sampling vegetation. Each diagram shows a possible situation relating to boundaries and mosaics: (a) sampling avoiding an obvious boundary; (b) sampling a homogeneous transition; (c) sampling avoiding a complex boundary; (d) sampling a mosaic.

another begins. In such cases, it is helpful to pick out first those areas which are clearly different from one another in their floristics and physiognomy. These may be quite small patches, but if it is possible to recognise the 'black and white' areas initially, then indeterminate 'grey' zones between should become apparent. One practical way to do this is to choose one more or less homogeneous tract of vegetation and then keep walking, first in one direction, then repeating the procedure in another, until clear differences are perceived, backtracking each time to see where and how the changes occur.

Ecotones are directional sequences of different vegetation types which are often clearly related to environmental gradients, such as the height of the water table in fens and swamps around a lake margin or the amount of salt-spray deposited over the top of an exposed sea-cliff. They have great ecological interest but, where the gradation is diffuse, they can pose an acute example of the kind of problem discussed above (see also Figure 2), with vegetation types giving way indistinctly one to another. Here, it may be necessary to suspend judgement about where one plant community gives way to another until after sampling: then, taking quadrats may be best done by using a transect at right angles to the ecotone and sampling at regular intervals along it.

2.4. Mosaics

Mosaics can present similar difficulties to ecotones, though they are frequently non-directional, with complex patchworks of different vegetation types intermingled over ground that has some kind of environmental patterning (Figure 2). But there is an additional problem here, because many types of vegetation which occur in more or less uniform stands have mosaics within them on a finer scale that is related to the morphology of the different species, such as bulky tussock grasses or clonal plants like *Cirsium acaule* or *Mercurialis perennis*, or to the localised impact of environmental factors, such as cattle dung in pastures, the persistence of which gives rise to 'avoidance-mosaics' of less heavily-grazed patches of herbage.

It is often difficult to decide whether patterning in vegetation is of this kind or is really a mosaic of two different vegetation types. A handy rule of thumb is to see whether such patterns involve qualitative differences in species in the different components of the pattern, or just quantitative variation among species which are present throughout the mosaic.

If different species are present within the patches of the elements in the pattern, it is better to delimit stands of different vegetation types and to sample accordingly. If there are simply local variations in abundance, then the patterning can be treated as part of the variation within a single stand.

Where extensive areas are covered by a mosaic, where the pattern is repeatedly encountered in the same form or where the scale of the mosaic is such that it is impossible to lay down separate samples of the relevant size in one or more of the components, the mosaics should be sampled in their entirety (see Figure 2).

2.5. Sampling around threatened plant species

For conserving rare or threatened plant species, it is vital to recognise and understand the vegetation types in which they occur. Many such species occur in a variety of different plant communities: knowing what these are and how the vegetation types relate to differences in habitat is essential for successful conservation and species-recovery programmes.

Sampling the vegetation in the usual fashion around such threatened plants, provided the individuals occur in homogeneous stands, is a quite straightforward procedure and yields much more valuable data than informal lists of species associated around a rare plant. Where the plants are found on boundaries between vegetation types or in gradual transitions, like the bog-orchid *Hammarbya paludosa*, which typically occurs around the very edge of bog pools, the constituents of the mosaic can be sampled in the usual way and the distribution of the plants recorded in the descriptive notes.

2.6. Systematic or random sample location

The NVC style of recording is compatible with systematic, random or restricted randomised sample location, provided such samples do not fall across obvious boundaries between vegetation types and thus conflate information from different plant communities. Using such systematic or random sample arrays can be invaluable in the study of small-scale variation but it should be remembered that, in primary survey, they are often less economical than a strategy based on the location of samples by choice within homogeneous stands. They yield data dominated by the commoner vegetation types in an area and tend to under-represent or miss minor elements or small stands. This can be of critical importance in, say, sampling flushes on upland hillsides or bog pools on a mire surface.

2.7. Standardised cards for NVC sampling

It is very helpful to use standardised record sheets or cards for NVC sampling. These serve as a prompt to ensure that all relevant information is recorded and can greatly assist data coding and analysis. The kind of card used in the original NVC survey is shown in Figure 3.

		NVC record sheet	
Location	**Grid reference**	**Region**	**Author**
Site and vegetation description		**Date**	**Sample no.**
		Altitude m	**Slope** °
		Aspect °	**Soil depth** cm
		Stand area m x m	**Sample area** m x m
		Layers: mean height m m cm mm	
		Layers cover % % % %	
		Geology	
Species list		**Soil profile**	

Figure 3   A blank NVC sample card.

3 Choosing the size and shape of samples

3.1. Sample sizes

A relatively small number of sample sizes is sufficient for sampling the range of vegetation found in Britain (see Figure 4). In the NVC, these are scaled, not by any detailed calculation of minimal area, but by experienced assessment of appropriateness to the range of structural variation found among our plant communities. These sample sizes are:

2x2 m	short herbaceous vegetation dwarf-shrub heaths
4x4 m	short woodland field layers tall herbaceous vegetation heaths open vegetation
10x10 m	dense scrub tall woodland field layers species-poor herbaceous vegetation
50x50 m	woodland canopy and shrub layers sparse scrub

Sample sizes for hedges, banks, verges and water margins are discussed in sections 3.5-3.6. In existing surveys of fine-grained grasslands, samples of 1x1 m will often be found adequate for comparison with NVC data. Likewise, somewhat differently sized samples of woodland will often prove satisfactory (see Hall *et al.* 2004).

3.2. Sample shapes

In general, samples should be square, but sometimes this is impossible. Zones of vegetation can be very narrow, as in tight sequences on a cliff top, around a fluctuating pond or along the edge of a salt-marsh creek. In other cases, stands can be very irregular, as with sinuous bog pools. In such situations, alternative-shaped samples of the appropriate area should be used (Figure 5, see also Plate 3).

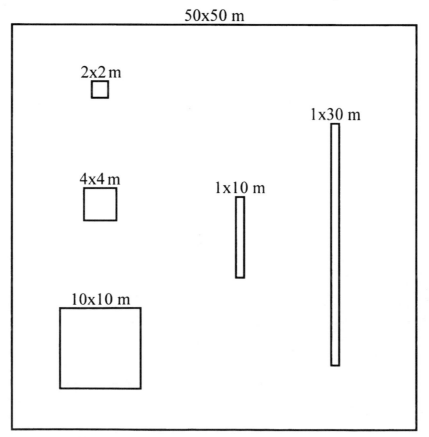

Figure 4   Comparative sizes of NVC samples. The diagram shows different quadrat sizes as used for different types of vegetation (see Section 3.1). The rectangular shapes apply to hedges (see Section 3.5).

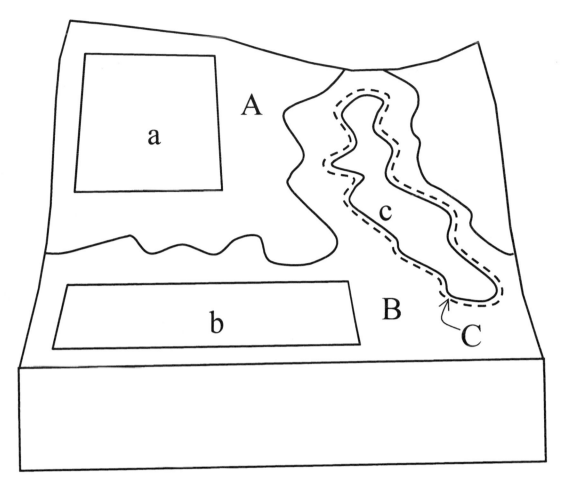

Figure 5   Sampling using rectangular quadrats or irregular shapes. The diagram shows sampling from bog hummock and hollow vegetation. Three homogeneous stands of mire vegetation (A-C) have been distinguished and a sample plot laid out in each: (a) 2x2 m where possible on the hummock; (b) with an identical area of different shape in the hollow or the entire stand; and (c) in the small pool (redrawn from Rodwell 1991b).

3.3. Small stands of vegetation

Certain vegetation types, like those in some flushes, salt-marsh pans or on cliff-ledges, often occur as stands smaller than the appropriate sample size. The stands should then be sampled in their entirety (see Plate 4). With the vegetation of small crevices on rock exposures, it is sensible to record a sample of 2x2 m or 4x4 m which includes a number of vegetated areas.

3.4. Sampling in woodlands

Sampling woodland vegetation poses some particular problems. The range of size among the plants represented is very great, from hepatics, mosses and lichens that can be tiny, through herbs and ferns of middling size, to shrubs and sometimes enormous trees. The eye needs to accommodate to these different scales of floristic and physiognomic variation within the layers, but detecting homogeneity among trees and shrubs is not actually all that difficult: it can be much easier than in some grasslands or mires, for example, where the

patterning is very fine. However, a lot of ground might have to be covered to assess the extent of diversity within a single tract of woodland canopy and understorey, and individual samples need to be large enough to represent adequately the scale of uniformity in the selected stands. Also, beneath such a homogeneous stretch of shrubs and trees, the field and ground layers may be much less coarsely structured. In the NVC, a large plot of 50x50 m was used for the canopy and understorey, and within this either a 4x4 m or 10x10 m plot for the field and ground layers, depending on their own scale of organisation, the records then being combined to constitute a single sample. Where there is variation in the field and ground layers within a 50x50 m canopy and understorey sample, more than one 4x4 m or 10x10 m sample can be taken, each then being combined with the same larger sample (see Figure 6). Other aspects related to sampling in woodlands are covered by Hall *et al.* (2004).

This approach to the sampling and description of woodland vegetation is different from that used in previous schemes. In some of these, notably the

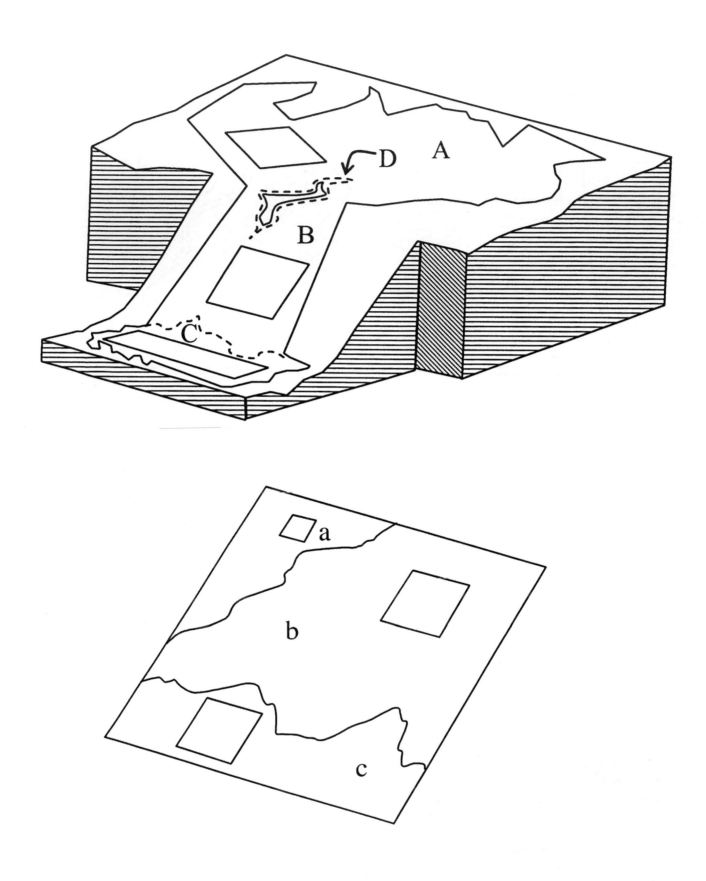

Figure 6   Sampling in woodlands. In the diagram, four homogeneous stands of canopy and understorey have been distinguished and a sample plot laid out in each, 50x50 m where possible (A and B) or an identical area of different shape (C) or, with a small patch, the entire stand (D). In each plot, homogeneous areas of field and ground vegetation are delineated, as in B enlarged beneath, with samples of either: (a) 4x4 m; or (b and c) 10x10 m (redrawn from Rodwell 1991a).

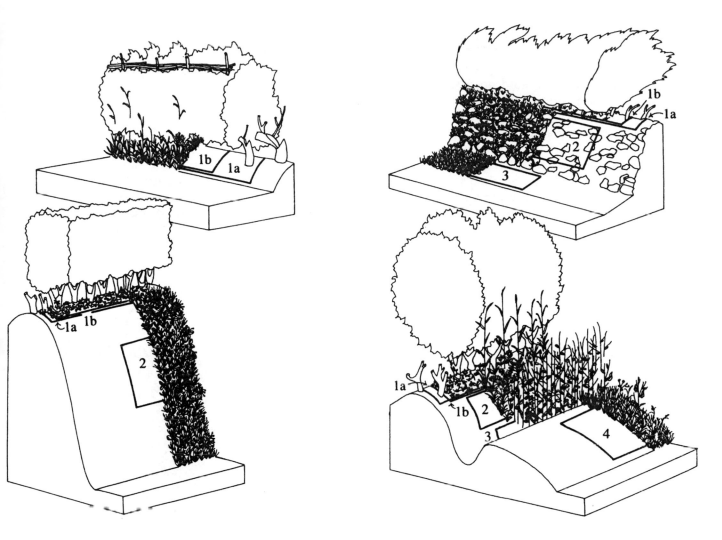

Figure 7   Sampling hedges, banks and verges. Samples are taken from homogeneous stretches of the hedgerow core, with trees and shrubs recorded in a 30 m strip (1a in each example), and of the associated field and ground vegetation, recording in a 10 m strip (1b). Vegetation on adjacent banks, ditches and verges is separately recorded in samples of relevant size (2, 3, 4) (reproduced from Rodwell 1991a).

Stand Type approach of Peterken (1981) and classification of Rackham (1980), attention has been focused on variation among the trees and shrubs, with information from the field and ground layers being used for subsequent refinement or corroboration of the classification. In Bunce (1982), by contrast, recording of vascular plants and bryophytes provided the main basis for the classification of Plot Types, with the additional difference that samples were located randomly and therefore were frequently heterogeneous in their composition. The implications of these differences for classifying and understanding British woodlands are dealt with in further detail in *British Plant Communities, Volume 1: Woodlands and Scrub* (Rodwell 1991a), the NCC *Woodland Survey Handbook* (Kirby 1988), and the updated volumes of Peterken (1993) and Rackham (2003).

3.5. Sampling hedges, banks and verges

Hedges are treated in the NVC as linear woodlands with canopy and field/ground layers being sampled separately. To integrate with existing protocols for hedgerow survey (Hooper 1970), a strip 30 m long is used for recording trees and shrubs, selected as a representative length. Within this, the field and ground layer are recorded in a uniform 10 m strip, 1 m wide or the width of the herbaceous zone in the hedge if narrower. Associated vegetation of banks, verges and ditches is then recorded separately (see Figure 7) and notes are made on the way the elements of the habitat are related.

3.6. Sampling water margins

Where sufficiently large zones or patches of fen and swamp vegetation occur around open waters, these should be sampled in the usual fashion, using 4x4 m or 10x10 m quadrats or rectangular equivalents as appropriate. However, around the edges of many ponds and lakes or along the banks of rivers, streams or canals, such zonations are often very condensed or fragmentary (see Figure 8). Smaller samples from each of the elements then have to be taken. What is important is not to combine records from what are actually different vegetation types. Also, it is generally better to regard any associated submerged or floating aquatics as structurally independent elements of the vegetation and record these separately (see Plate 5).

Other approaches to sampling emergent and aquatic vegetation of linear open waters have used ditch, stream or river lengths for recording lists of species with cover abundance values. NCC surveys of the Pevensey Levels (Glading 1980), North Kent Marshes (Charman 1981), Broadland (Reid *et al.* 1989, Doarks 1990, Doarks and Storer 1990, Doarks and Leach 1990, Doarks *et al.* 1990), North Norfolk (Leach and

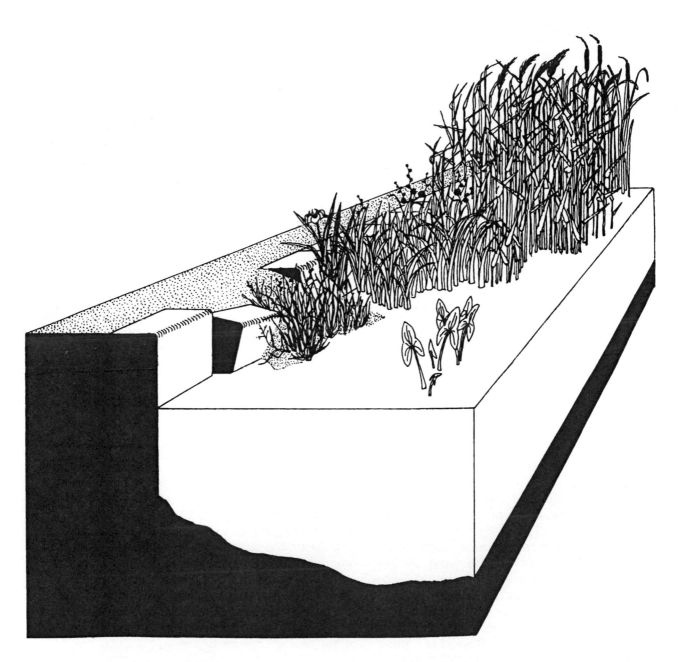

Figure 8   Sampling the margins of a canal. The diagram shows a typical pattern of patchy swamp vegetation along a disused canal. Each stand would be sampled separately using the NVC approach (reproduced from Rodwell 1995).

Reid 1989), the Derwent Ings (Birkinshaw 1991), Essex and Suffolk (Leach and Doarks 1991) and Devon (Leach *et al.* 1991) have all derived classifications of ditch lengths. Typologies of rivers have also been produced by Holmes *et al.* 1999 (see also Holmes *et al.* 1972, Holmes and Whitton 1977a, b, Holmes 1983) and Haslam (1978, 1982; see also Haslam and Wolseley 1981). Because of their different approaches to sampling and description of the vegetation, it is not always possible to relate such classifications to the NVC. However, used together, such techniques can yield complementary insights into variation in plant communities along different kinds of moving waters.

3.7. Sampling in open waters

Sampling the aquatic vegetation of open waters is difficult and can be dangerous. Only diving is entirely adequate for sampling submerged aquatics and then it is often possible to record species in the usual NVC fashion, sampling homogeneous stands using quadrats of appropriate size (Spence 1964). In fact, sampling is usually done by observations from a boat, from the bank or by wading into the shallows. Provided it is possible to delimit homogeneous stands of aquatic vegetation in this way and secure adequate records of all species present, such sampling is quite acceptable. Again, what is

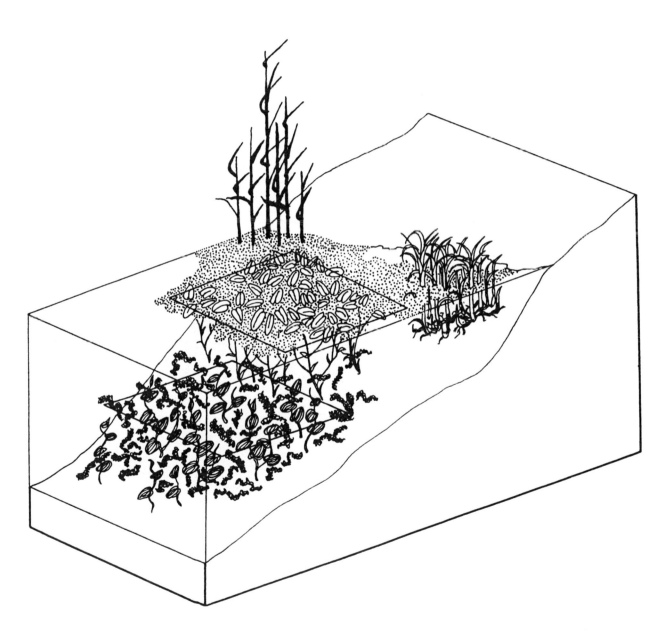

Figure 9   Sampling submerged and floating aquatics. The diagram shows sampling from superimposed layers of floating and submerged aquatic vegetation among emergents at a lake edge (reproduced from Rodwell 1995).

important is not to mix records from what are actually different vegetation types, all too easy with adventurous sampling by grapnel.

Another common problem is to ensure that different layers of aquatic vegetation, where submerged, floating-leaved or free-floating plants occur superimposed over one another, are sampled separately and without any emergents beneath and among which they are growing (see Figure 9 and Plate 5). This may seem an awkward and fussy recording procedure, but it has been the traditional phytosociological approach, reflecting the view that such assemblages of aquatics are distinct communities, related to different environmental conditions in particular sites and playing different roles in the successional colonisation of open waters. In practice, it can be difficult to devise a neat solution to such complexities: a pragmatic approach has to be adopted and, where especially dense mixtures of aquatics and emergents occur, there is always an opportunity to record the detailed structure of the vegetation and relationships between the various assemblages on the sample card.

Alternative approaches to sampling aquatic vegetation have sometimes combined such submerged and floating assemblages with emergents, in cross-sections or lengths of ditches and rivers. In other cases, aquatics have been sampled together from entire waterbodies or sub-sites within such lakes, and NCC developed a classification based on such data (Palmer 1989, 1992, Palmer *et al.* 1992). Although it can be difficult to relate such schemes in detail to the NVC, using such approaches together can help understand how detailed vegetation patterns are distributed among different site types.

4 Recording sample location and time

4.1. Grid reference

An eight-figure grid reference (or ten if possible) should be given for each sample, using the 100 km grid square numbers or letters, plus the usual eastings and northings (see Figure 10 a-f for examples). Samples in Northern Ireland should be located using the revised Irish National Grid now accepted as normative for both the Province and the Republic.

4.2. Latitude and longitude

Latitude and longitude have rarely been used to denote spatial location in Britain but provide the usual method elsewhere in Europe, so international projects may need to employ this technique.

4.3. Global Positioning Systems

In featureless terrain, as on extensive blanket bogs, summit heaths or salt-marshes, satellite-related Global Positioning Systems can be used to determine the location of samples. More sophisticated equipment can give an accuracy of 10 m or less.

4.4. Site name

A site name and administrative county should be given for each sample. Where a site is a field, farm or road, it should be supplemented by the name of the nearest settlement or large-scale natural feature.

4.5. Sample size and context

Where a number of samples are taken from a particularly varied or complex site, the records should be supplemented by notes on the spatial interrelationships of the samples, with maps where appropriate.

4.6. Date of sampling

The date of sampling should be recorded for each sample. With most vegetation types, accurate recording of species data will be easier during the flowering season and, in certain plant communities, some species will have disappeared by summer, as with *Hyacinthoides non-scripta, Anemone nemorosa* and *Allium ursinum* in woodlands (Kirby 1988). Even if old leaves or flower spikes of such vernal herbs persist, it is often impossible to estimate cover accurately after late spring. Where this is likely to affect identification of vegetation types, a note is given in *British Plant Communities*. In sampling meadows, it is best to time sampling before the hay cut, although care should then be taken not to damage the crop by trampling: provided grasses can be identified vegetatively, sampling the aftermath in September or early October may be adequate.

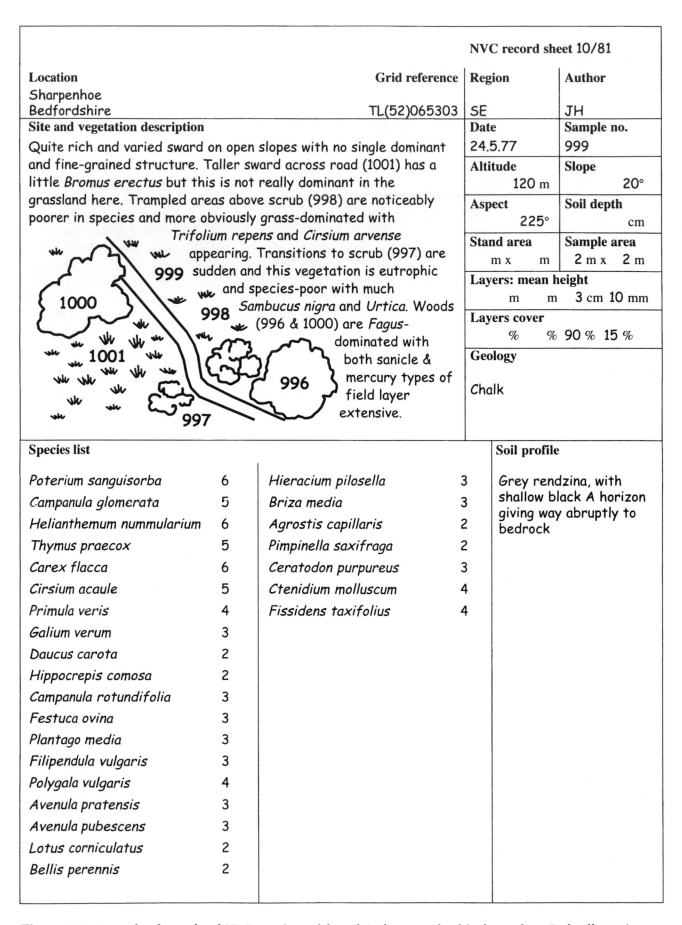

NVC record sheet 10/81

| Location Sharpenhoe Bedfordshire | Grid reference TL(52)065303 | Region SE | Author JH |

Site and vegetation description

Quite rich and varied sward on open slopes with no single dominant and fine-grained structure. Taller sward across road (1001) has a little *Bromus erectus* but this is not really dominant in the grassland here. Trampled areas above scrub (998) are noticeably poorer in species and more obviously grass-dominated with *Trifolium repens* and *Cirsium arvense* appearing. Transitions to scrub (997) are sudden and this vegetation is eutrophic and species-poor with much *Sambucus nigra* and *Urtica*. Woods (996 & 1000) are *Fagus*-dominated with both sanicle & mercury types of field layer extensive.

Date	Sample no.
24.5.77	999

Altitude	Slope
120 m	20°

Aspect	Soil depth
225°	cm

Stand area	Sample area
m x m	2 m x 2 m

Layers: mean height
m   m   3 cm 10 mm

Layers cover
%   % 90 % 15 %

Geology

Chalk

Species list

Poterium sanguisorba	6	Hieracium pilosella	3
Campanula glomerata	5	Briza media	3
Helianthemum nummularium	6	Agrostis capillaris	2
Thymus praecox	5	Pimpinella saxifraga	2
Carex flacca	6	Ceratodon purpureus	3
Cirsium acaule	5	Ctenidium molluscum	4
Primula veris	4	Fissidens taxifolius	4
Galium verum	3		
Daucus carota	2		
Hippocrepis comosa	2		
Campanula rotundifolia	3		
Festuca ovina	3		
Plantago media	3		
Filipendula vulgaris	3		
Polygala vulgaris	4		
Avenula pratensis	3		
Avenula pubescens	3		
Lotus corniculatus	2		
Bellis perennis	2		

Soil profile

Grey rendzina, with shallow black A horizon giving way abruptly to bedrock

Figure 10a Example of completed NVC sample card for calcicolous grassland (redrawn from Rodwell 1992).

Location	Grid reference	Region	Author
Linbrigg Northumberland	NT(36)895068	NE	CDP/JR

Site and vegetation description		

Tall-herb vegetation in patches over steeply-sloping cliff-base and tumbled debris, among open *Fraxinus*, *Corylus* and *Prunus padus* (see sample B39 for denser woodland vegetation). Dominated by large & luxuriant tussocks of *Luzula* and stocks of *Dryopteris* with scattered clumps of *Arrhenatherum* & other grasses, some *Pteridium* and tall dicotyledons with emergent inflorescences. Smaller herbs appear in more open areas and around margins, with patches of mosses over stow-bases and on litter which is quite thick in places. An impoverished version of this vegetation, mainly *Luzula* & ferns, turns up tunnels between rocky knolls, giving way to species-rich *Festuca-Agrostis* where grazed.

Date 22.6.1978	**Sample no.** 840
Altitude 210 m	**Slope** 45°
Aspect 320°	**Soil depth** cm
Stand area 3 m x 4 m	**Sample area** 2 m x 2 m
Layers: mean height m m 40 cm 10 mm	
Layers cover % % 95 % 15 %	
Geology Andesite lava	

Species list				Soil profile
Arrhenatherum elatius	3	*Cystopteris fragilis*	1	Brown earth, in pockets over irregular cliff-base, highly humic above, mid-brown sandy silt-loam below
Dryopteris filix-mas	6	*Vicia sepium*	2	
Luzula sylvatica	6	*Viola tricolor*	2	pH 0-5 cm = 5.0
Filipendula ulmaria	2	*Calliergon onspidatum*	1	
Angelica sylvestris	2	*Plagiothecium denticulatum*	2	
Geranium robertianum	3	*Mnium hornum*	2	
Holcus lanatus	4	*Rhizomnium punctatum*	1	
Pteridium aquilinum	5	*Brachythecium rutabulum*	4	
Mercurialis perennis	2	*Eurhynchium praelongum*	1	
Dactylis glomerata	2	*Eurhynchium swarzii*	1	
Urtica dioica	2	*Lophocolea bidentata s.l.*	2	
Hyacinthoides non-scripta	2			
Geum rivale	2			
Primula vulgaris	2			
Silene dioica	2			
Stellaris holostea	2			
Valeriana officinalis	1			
Cardamine pratensis	1			
Conopodium majus	1			

Figure 10b Example of completed NVC sample card for mesotrophic grassland (redrawn from Rodwell 1992).

Location Cauldrus, Yesnaby Orkney	Grid reference HY(130)223166	Region Scotland	Author AJCM

Site and vegetation description

Dwarfed patchy cover of wind-pruned & close-grazed *Calluna* with some *Empetrum*, forming irregular hummocks with much bars & stony ground between on more exposed stretches of cliff-top ablation surface (1). In other places, as here, a little grassier though still species-poor and with just sparse & grazed-down associates, apart from plantains which are locally plentiful. Sub-shrubs thin out quickly in plantain-rich sward to seaward (2), with open *Festuca-Arenaria* turf on eroding cliff-edge (3). Wet areas of flat ground throughout have a sort of salt-marsh vegetation with much *A. stolonifera* & *Glaux*. This general sequence

(4)   sample 980   continues southwards with

2 with patches   some variation in

3   of 4   **1**   proportions & clarity of the zones. In places there is a transition landwards to non-maritime heath but around the settlements enclosure & improvements have often converted cliff-top vegetation to pasture.

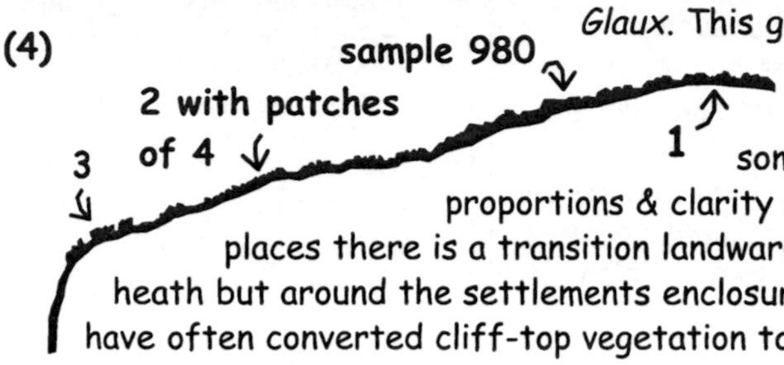

Date	Sample no.
2.7.1976	980

Altitude	Slope
24 m	5°

Aspect	Soil depth
230°	24 cm

Stand area	Sample area
50 m x 100 m	2 m x 2 m

Layers: mean height

m   m   4 cm   mm

Layers cover

%   %100 %   %

Geology

Devonian Old Red
Sandstone flags and shales

Species list

Calluna vulgaris	6
Empetrum nigrum nigrum	5
Festuca rubra	5
Plantago lanceolata	4
Plantago maritima	5
Thymus praecox	2
Agrostis stolonifera	2
Lotus corniculatus	5
Scilla verna	2
Cerastium fontanum	1
Carex panicea	4
Euphrasia officinalis	2
Trifolium repens	2

Soil profile

Humic ranker, very sandy below

pH 0-5 cm = 5.1

Figure 10c Example of completed NVC sample card for heath vegetation (redrawn from Rodwell 1991b).

Location Beinn Dearg Ross	Grid reference NH(28)256816	Region SCOT	Author DAR & DM

Site and vegetation description		Date 1962	Sample no. R56052

Open *Juncus-Festuca* vegetation forming irregular patches in centre of stone polygons over high summit ablation terraces with distinctive *Carex-Racomitrium* heath rich in cushion herbs around, then network of unvegetated blocks. *Gymnomitrium* locally prominent in a patchy crust & *Salix herbacea* present in this vegetation nearby but not in quadrat. This open cover passes on lower slopes to more extensive *Carex-Racomitrium* heath, still often with plentiful cushion herbs.

↓ gentle slope downhill

Altitude 990 m	Slope 0°
Aspect -°	Soil depth ‹25 cm
Stand area 3 m x 4 m	Sample area 2 m x 2 m

Layers: mean height
m   m   5 cm 10 mm

Layers cover
%   % 40 % 25 %

Geology

Moine siliceous granulites

Species list

Deschampsia flexuosa	3
Festuca ovina	3
Carex bigelowii	2
Juncus trifidus	4
Luzula spicata	2
Alchemilla alpina	4
Armeria maritima	3
Minuartia sedoides	3
Omalotheca supina	2
Sibbaldia procumbens	2
Silene acaulis	3
Polytrichum piliferum	2
Racomitrium heterostichum	3
Racomitrium lanuginosum	5
Gymnomitrium concinnatum	2
Cornicularia aculeata	2
Cetraria islandica	3
Cladonia pyxidata	3
Cladonia uncialis	2

Soil profile

Ranker, evidently a truncated podzol with solifluction

pH 0-5 cm = 4.9

Figure 10d Example of completed NVC sample card for calcifugous grassland and montane vegetation (redrawn from Rodwell 1992).

Location Big Water of Fleet Galloway	Grid reference NX(25)559643	Region Scotland	Author KH

Site and vegetation description

Vegetation of this kind occupies the very gently-sloping water-backs running down to and along-side the river flats, gathering drainage from the complex of *Scirpus-Erica* wet heath and *Molinia*-dominated mires on the hills around. Molinia is strongly-tussocky with scattered *Myrica* bushes on and between the clumps, locally dense *Juncus* & herbaceous associated between with a patchy carpet of bryophytes between culms & litter. Surrounding slopes quite heavily grazed by sheep & cattle but only patchy predation here. Margins however grade to a close-cropped flushed *Festuca-Agrostis* sward with rushes on alluvial flats.

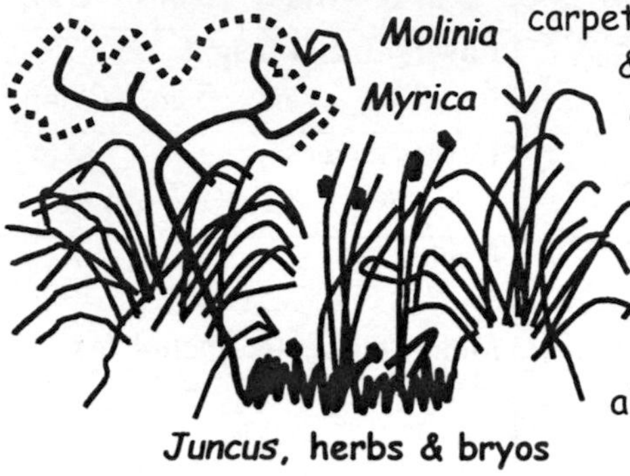

Date 12.6.1978	**Sample no.** 76
Altitude 105 m	**Slope** 0°
Aspect -°	**Soil depth** 39 cm
Stand area m x m	**Sample area** 2 m x 2 m

Layers: mean height
m   m 45 cm 80 mm

Layers cover
%   % 60 % 60 %

Geology

Fluvial gravels

Species list

Molinia caerulea	7	Sphagnum papillosum	5	
Juncus acutiflorus	6	Aulacomnium palustre	1	
Myrica gale	5	Sphagnum tenellum	4	
Succisa pratensis	4	Rhytidiadelphus squarrosus	1	
Viola palustris	4	Sphagnum capillifolium	2	
Carum verticillatum	2	Odontoschisma sphagni	1	
Cirsium palustre	1	Thuidium tamariscinum	2	
Scirpus cespitosus	2			
Luzula multiflora	2	litter (mostly *Molinia*)	8	
Galium palustre	1			
Valeriana officinalis	1			
Angelica sylvestris	2			
Narthecium ossifragum	1			
Galium saxatile	1			
Drosera rotundifolia	1			
Carex panicea	1			
Festuca ovina	2			
Carex echinata	1			
Anthoxanthum odoratum	2			

Soil profile

Peat, rather fibrous & uncompacted

pH 0-5 cm = 4.7

Figure 10e Example of completed NVC sample card for mire vegetation (redrawn from Rodwell 1991b).

Plate 1   Before vegetation can be sampled using the NVC, homogeneous stands need to be identified by eye based on the patterned arrangement of species over the ground and vertical layering. This is revealed in the view above, for example, by the consistency and repetition of colour, texture and structure of the vegetation: the homogeneous stand in the foreground is composed mainly of bog-myrtle *Myrica gale*, heather *Calluna vulgaris*, and purple moor-grass *Molinia caerulea* (M15 *Scirpus cespitosus–Erica tetralix* wet heath community). © JNCC

Plate 2   Boundaries between vegetation communities can take a variety of forms and be more or less discrete. This is illustrated in the view above by, for example, the patterning of the white inflorescences of *Eriophorum angustifolium* (M1 *Sphagnum auriculatum* bog pool and M17 *Scirpus cespitosus–Eriophorum vaginatum* blanket mire communities, Mamores, Scottish Highlands). © JNCC

Plate 3   Linear features can pose particular problems for sampling, especially if they are narrower than the standard quadrat size recommended in the NVC. The wall crevices and road verge shown here would be best sampled using a 1x4 m quadrat, instead of the normal 2x2 m size. The verge has MG1 *Arrhenatherum elatius* grassland and the wall OV39 *Asplenium trichomanes–Asplenium ruta-muraria* communities. © JNCC

Plate 4   Awkwardly-shaped and small stands of vegetation, as illustrated by this patch of eared willow *Salix aurita* scrub growing in a gully in Glen Nevis, Scottish Highlands, pose another problem for recording in the NVC. In such situations, the stand should be sampled in its entirety. © JNCC

Plate 5   When applying the NVC methodology to aquatic vegetation, the layers of floating, submerged and emergent plants should be sampled as separate layers, as these are regarded as separate communities. In this case, the white water-lily *Nymphaea alba* (A7 *Nymphaea alba* community) would, for example, be recorded separately from the bottle sedge *Carex rostrata* and bogbean *Menyanthes trifoliata* (S9 *Carex rostrata* swamp). © JNCC

Plate 6   It is important to record all vegetation species present in samples, including bryophytes and macrolichens. For example, the relative frequency and abundance of woolly-hair moss *Racomitrium lanuginosum* and lichens such as *Cladonia portentosa*, seen here with crowberry *Empetrum nigrum hermaphroditum*, are important for separating the related montane vegetation types H20 *Vaccinium–Racomitrium* heath and H19 *Vaccinium–Cladonia* heath. © JNCC

Plate 7   This woodland stand is dominated by hazel *Corylus avellana* in the canopy, and by ramsons *Allium ursinum*, ferns and bryophytes, on the ground. Nevertheless, it conforms most closely to the W9 *Fraxinus excelsior–Sorbus aucuparia–Mercurialis perennis* community in the NVC. This illustrates that the species used to name the NVC community types are not always prominent or even present in any one stand. © JNCC

Plate 8   The NVC has been trialled widely and discussed by many individuals. Here, users are being instructed on the identification and recording of vegetation mosaics in the bog pools and mire plane of the New Forest, southern England (M21 *Narthecium ossifragum–Sphagnum papillosum* valley mire and M1 *Sphagnum auriculatum* bog pool communities). © John Rodwell

Location	Grid reference	Region	Author
Point Burn Woods Durham	NZ(45)145565	NE	JR

Site and vegetation description

Irregular-shaped stand of wet woodland below a flush line with uneven topped & somewhat open cover of Alnus, many suckering, with occasional *Fraxinus & Quercus* and, a little lower, scattered sickly *Betula*. Understorey quite dense with *Corylus* predominating as large uncoppiced bushes and quite numerous saplings up to 3-4 m, then below a scattered cover of *Ribes, Rosa, Salix* and *Viburnum*. Field layer as here in central wettest areas with patchy dominance of *Juncus, Carices & Equisetum hyemale*. Grades around to drier *Deschampsia cespitosa* field layer wider same canopy (sample 544), then to oak-birch woodland with *Holcus mollis* field layer (546). Patches of drier ground around planted with *Fagus* & occasional conifers. No obvious signs of grazing or other recent treatments.

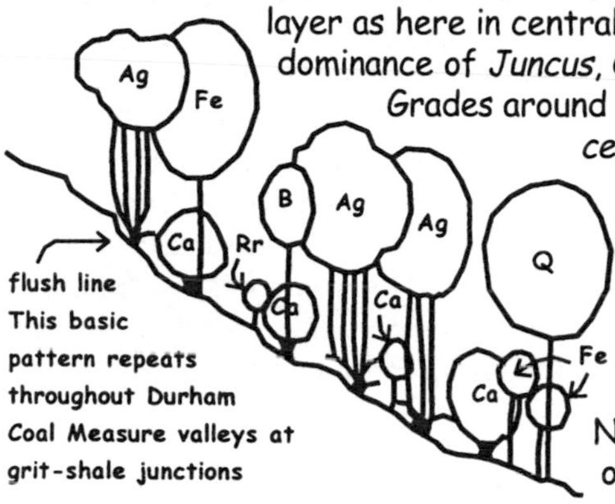

flush line
This basic pattern repeats throughout Durham Coal Measure valleys at grit-shale junctions

Date	Sample no.
3.8.1977	543

Altitude	Slope
60 m	10°

Aspect	Soil depth
270°	70 cm

Stand area	Sample area
30 m x 70 m	30 m x 70 m
5 m x 7 m	4 m x 4 m

Layers: mean height
85 m 25 m 95 cm 20 mm

Layers cover
15 %   3 % 70 % 25 %

Geology

Carboniferous coal measure sandstones & shales

Species list

Canopy in 30 x 70 m

Alnus glutinosa	7
Betula pubescens	3
Fraxinus excelsior	6
Betula pendula	2
Quercus petraea	1
Quercus hybrid	4

Understorey in 30m x 30 m

Corylus avellana	5
Alnus glutinosa sapling	2
Crataegus monogyna	2
Fraxinus excelsior sapling	2
Acer pseudoplatanus sapling	1
Prunus padus	2
Rosa canina agg.	1
Rubus idaeus	1
Quercus robur sapling	1
Ribes rubrum	1
Salix cinerea	1
Viburnum opulus	1

Field layer in 4 x 4 m

Lonicera periclymenum	3
Equisetum sylvaticum	4
Athyrium filix-femina	5
Crispum palustre	1
Equisetum hyemale	1

Filipendula ulmaria	5
Juncus effusus	6
Allium ursinum	4
Viola riviniana	1
Lysimachia nemorum	3
Galium aparine	1
Angelica sylvestris	1
Ribes rubrum seedling	1
Ranunculus repens	1
Deschampsia cespitosa	5
Bromus ramosus	2
Carex remota	4
Carex laevigata	3
Festuca gigantea	1

Ground layer in 4 x 4 m

Mnium hornum	1
Plagiomnium rostratum	1
Plagiomnium undulatum	1
Eurhynchium praelongum	5
Lophocolea bidentata	2
Cirriphyllum piliferum	1
Brachythecium rutabulum	1
Eurhynchium striatum	3
Thuidium tamariscinum	1
bare soil & litter	3

Soil profile

Stagnogley, with deep upper horizon of wet rather structureless silt in centre of flush, becoming clayey below with shale fragments, whole profile tending to slump downslope

pH 0.5 cm = 5.7

Figure 10f Example of completed NVC sample card for woodland.

5 Recording vegetation data

5.1. The species list

The NVC demands that all vascular plants, bryophytes and macrolichens (*sensu* Dahl 1968) rooted or attached within the sample should be accurately identified and listed (see Plate 6). Critical taxa should be treated in as much detail as possible and all doubtful material checked by referee or catalogued as vouchers. Where compromises are made in identification, as with microspecies of *Rubus fruticosus* agg., for example, or the various *Taraxacum* spp., this should always be noted in the survey documentation.

In woodlands, where vegetation is conspicuously layered, species lists should be made separately for the layers, and species occurring in more than one layer recorded for each, as canopy trees, shrubs or saplings and seedlings. Such detail can be invaluable for understanding the state of regeneration. In the same way, climbers and lianes can also figure in several layers of woodland and scrub vegetation.

Where species occupy distinctive niches in a fine mosaic that is included within a more coarsely heterogeneous sample, as with boulder and crevice bryophytes in acidophilous woodland for example, this can be noted within the species list and explained in the accompanying text.

Algae were generally not recorded in the NVC except where they were structurally very important among assemblages of vascular plants. In some salt-marsh swards, for example, ecads of *Fucus vesiculosus* were prominent, and among certain aquatic communities, *Chara* and *Nitella* were recorded to the genus.

Some subsequent surveys have extended this approach to recording the fungi occurring in NVC plant communities (Watling 1981, 1987) and it has been suggested that a phytosociological approach should be extended to marine algal assemblages. Some communities of epilithic and epiphytic lichens have already been characterised by James *et al.* (1977) and epiphytic bryophytes were included in a phytosociological woodland survey by Graham (1971).

5.2. Authorities for species names

The taxonomic authorities used in species lists for recording vegetation should always be specified in survey documentation. In the NVC, the authority for vascular plants was *Flora Europaea* (Tutin *et al.* 1964 *et seq.*), for bryophytes Corley and Hill (1981), and for macrolichens Dahl (1968). This last is now widely acknowledged as outdated and should be replaced by Purvis *et al.* (1992, 1994). The Checklist of bryophytes has also been updated (Blockeel and Long 1998). For recording vascular plants, many surveyors now use Stace (1997, 1999), and the VESPAN software (see section 7.2) provides an automatic cross-reference between Stace and *Flora Europaea* names.

5.3. The Domin scale of cover/abundance

For every species recorded in the sample, an estimate should be made of its quantitative contribution to the vegetation. Cover/abundance is a measure of the vertical projection on to the ground of the extent of the living parts of a species (see Figure 11). In the NVC, this is estimated using the Domin scale (*sensu* Dahl and Hadač 1941):

Cover	Domin
91–100%	10
76–90%	9
51–75%	8
34–50%	7
26–33%	6
11–25%	5
4–10%	4
<4% (many individuals)	3
<4% (several individuals)	2
<4% (few individuals)	1

Even within vegetation which is not very conspicuously layered, the total of all the Domin values for the species can exceed 100% cover because of structural overlap of the plants.

In practice, in herbaceous vegetation, it is usually sensible to record the cover/abundance of all vascular plants and ferns first, then the values for mosses, liverworts and lichens.

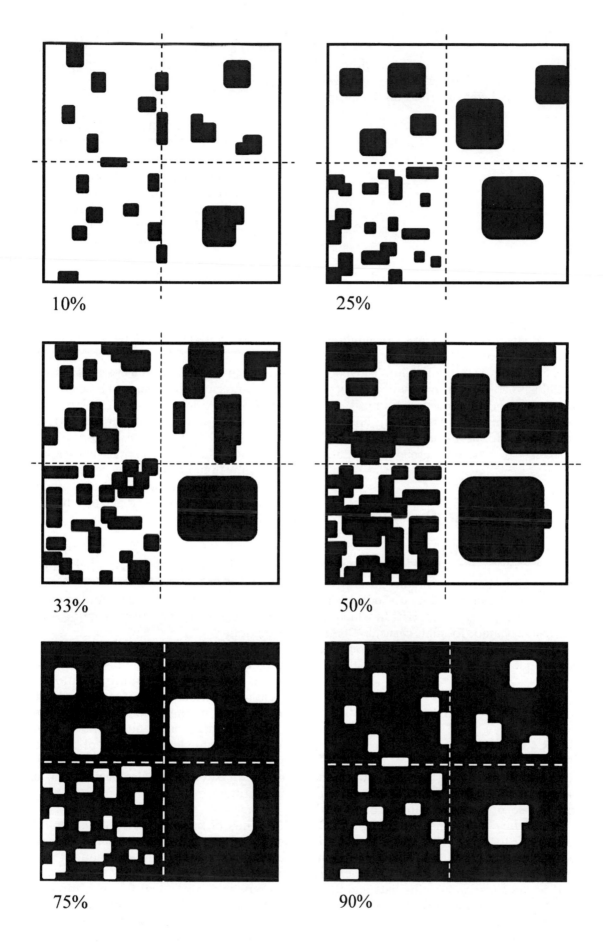

Figure 11   A visual interpretation of Domin cover/abundance thresholds. In the diagrams, each sub-square has the same total area of black: the top left diagram, for example, has 10% black in each sub-square.

5.4. Other cover scales

Occasionally in British surveys (e.g. Shimwell 1968, Wheeler 1975), and much more extensively in Europe, cover has been recorded on the 5- or 6-point Braun-Blanquet scale:

Cover	Braun-Blanquet scale
76–100%	5
51–75%	4
26–50%	3
6–25%	2
1–5%	1
<1%	+

Domin values can be converted to this scale without major discrepancies, but unambiguous conversion from Braun-Blanquet to Domin is impossible. For later interconversion between a variety of scales, recording in percentage covers or in 5% cover bands is most economical. The DAFOR scale (Dominant, Abundant, Frequent, Occasional, Rare) should never be used in NVC surveys because it has no agreed quantitative meaning.

5.5. Species outside the sample but in the stand

It is sometimes informative to record species that are absent from a sample but present in the homogeneous stand of vegetation within which the sample has been located. In such cases, a 'Domin' score of 11 can be used for recording and computer coding. Signs such as + or x should be avoided.

5.6. Using prepared species lists for recording

For surveys which are being carried out among limited ranges of vegetation types, it can be very economical to use pre-printed sample cards with prepared lists of the species that are thought likely to be encountered together with space to add any additional species that are found. The descriptions of vegetation types in *British Plant Communities* can be used to devise such species lists and these can serve as a training aid for sharpening identification skills, as with the list of key woodland bryophytes derived by Hall *et al.* (2004). As always, care should be taken that such lists do not condition the expectations of surveyors.

Where data are to be subsequently computer-coded, the number codes can be printed alongside species names on sample cards to save time when entering data.

5.7. Total vegetation cover

The total percentage cover of the vegetation should be recorded, broken down into any obvious layers. The standard NVC sample card provides four boxes for this and these should be used as systematically as possible for (from right to left) ground layer of cryptogams, herb or sub-shrub layer, shrub layer or woodland understorey and woodland canopy. Again, the total of the various layers may be well over 100% because of structural overlap.

5.8. Vegetation height

The mean height of the various layers of the vegetation listed above should be recorded: ground layer (in mm), field or sub-shrub layer (cm), shrub layer/understorey (m) and canopy (m). Again, the NVC sample card provides a series of boxes for these data. Many Continental surveys also record maximum height and this can be especially informative where flowering shoots greatly exceed the mean height of the herbage or where layers are very irregular in height.

5.9. Other structural details

This information should be supplemented by any useful details of structural complexities in the vegetation.

The NVC sample card provides considerable space for this and every opportunity should be taken to refine observational skills for understanding what can be seen in the field. Perceptive notes made while sampling are much more useful than summaries made later and data of this kind can be invaluable for interpreting not only present vegetation patterns but also past change and future possibilities.

The kinds of information recorded could include details of any patterns of spatial organisation or dominance among the component species; of contributions from especially prominent life-forms like winter annuals, mat-formers, rosette plants or tall-herbs; and any suggestions of phenological change through the growing season, or response to environmental shifts or management operations. Simple sketches can be very useful for summarising this sort of detail.

5.10. Zonations to neighbouring vegetation types

The spatial relationships between the stand being sampled and any neighbouring vegetation types should be described. Again, visual summaries such as maps or sectional drawings can be highly informative. This sort of information is especially necessary where NVC sampling forms part of a site survey and where complexes of different vegetation types make up a distinctive habitat like a lowland raised mire, sea-cliff or mosaic of woodland, scrub and grasslands with intricate margins.

5.11. Signs of succession

It is important to avoid jumping to conclusions about successional processes on the basis of spatial comparisons between NVC samples, but reliable indications of change can often be detected in the floristic and structural details of vegetation. Things to note could include variations in the vigour of species, the predominance of growth phases, the age structure of populations of individuals, or signs of senescence, death or regeneration.

6 Recording environmental data

6.1. Altitude

The altitude of samples can be estimated to the nearest 10 m from Ordnance Survey 1:50,000 maps or measured using a pressure altimeter.

6.2. Slope

The mean slope of the ground in samples can be measured to the nearest degree using a clinometer or level. Flat sites should be recorded as 0°.

6.3. Aspect

The mean aspect of samples can be measured to the nearest degree using a compass. North should be recorded as 360° and flat sites using a dash.

6.4. Geology

Details of bedrock and superficial geology should be given from Geological Survey maps and field observation.

6.5. Soil

Where soil details are being recorded, a small soil pit should be dug and the profile described using the horizon notation of the Soil Survey (Hodgson 1976). The profile should be allocated to one of the Avery (1980) soil groups. pH should be measured from a fresh sample of the superficial horizon using an electric meter in a 1:5 soil–water paste.

6.6. Recording bare rock, bare soil and litter

The extent of bare rock, exposed soil and litter should be recorded using the Domin scale or a percentage cover score.

6.7. Describing the terrain

These simple quantitative records should be supplemented by observations on the character of the landscape from which the sample has been taken: both coarse-scale features and elements of micro relief, signs of erosion or deposition, patterning among rock outcrops, talus slopes or stony soils; details of the drainage pattern down slopes and valleys, through and around mire systems and flushes, from snow-beds and springs; exposure to wind, salt-spray or frost. This kind of information, even if only qualitative, can help in interpreting the character of the vegetation and understanding its relationship to the habitat. Sketch-maps or profiles can greatly help in understanding the relationships between the vegetation and the terrain.

6.8. Recording the aquatic environment

Where possible, samples of aquatic or emergent vegetation should be supplemented by records of water depth, speed of flow, any evidence of periodic, seasonal or irregular fluctuations in level or flow, any signs of impact of waves or currents and the character of the bed, shore or bank. The extent of open water can be coded as a Domin value or percentage cover and conductivity and pH measured with electric meters.

6.9. Recording biotic impacts

Even where documentary or oral evidence is lacking, there are often signs in the vegetation and habitat themselves of biotic impacts, including treatments of the vegetation by man. Notes should be made on evidence of grazing or browsing by livestock or wild herbivores, trampling, dunging, mowing, burning, underwood or timber extraction, amenity use, and so on. Standard techniques for recording these consistently may be helpful; the Scottish Natural Heritage handbook for surveying land management impacts on upland habitats (MacDonald *et al.* 1998) offers a useful approach.

7 Characterising vegetation types

7.1. Making data tables by hand

It is possible to use individual samples of vegetation to identify a plant community by comparison with NVC data, but usually numerous samples are sorted and grouped to provide a summary of the floristics of each vegetation type in a survey.

Such sorting can be done by hand: before computational techniques were available, this was the normal method of making floristic tables and defining plant communities. However, even where there is access to computer facilities, hand-sorting of NVC samples can be invaluable in helping understand the principles of the floristic tables which are the basis of the definition of plant communities.

7.2. Computer analysis of NVC samples

NVC samples from vegetation surveys are now commonly sorted by some kind of multivariate analysis using computers; with speedy, capacious PCs, a technique such as TWINSPAN (Hill 1979) can analyse many hundreds of samples of complex vegetation in a matter of minutes. Packages like VESPAN III (Malloch 1999) and TURBOVEG (Hennekens and Schaminée 2001), with flexible facilities for data-editing, table-making and rearrangement, greatly assist the ancillary processes of refining, summarising and displaying the results of such analyses.

The ecological interpretation of the results remains the responsibility of the surveyor. All that computer analysis can do is define sample groups on the basis of statistical similarities and differences: to characterise vegetation types from the end-groups produced by such analysis requires skill and experience. For comparison with NVC data, a first step is to construct floristic tables that summarise the frequency and abundance values of the constituent species among the samples.

7.3. Frequency of species

The term frequency is used to describe how often a species is encountered in different stands or samples of a vegetation type, irrespective of how much of that species is present in each stand or sample. It is summarised in floristic tables using the Roman numerals I–V and referred to in descriptions of vegetation types using the terms listed on the right below:

Frequency
  class

I	= 1–20% (i.e. 1 stand in 5)	*scarce*
II	= 21–40%	*occasional*
III	= 41–60%	*frequent*
IV	= 61–80%	*constant*
V	= 81–100%	*constant*

7.4. Abundance of species

The term abundance is used to describe how much of a species is present in a stand or sample, irrespective of how frequently the species is encountered in moving from one stand to another. Floristic tables generally show the range of abundance for each species in a community using the Domin scale (see Section 5.3). In descriptions of vegetation types, abundance is referred to using terms such as 'dominant' (or 'prominent' or 'abundant' where there is high cover but no real dominance) and, for low covers, expressions such as 'sparse'.

7.5. NVC floristic tables

The floristic tables used to define the NVC vegetation types were the product of many rounds of multivariate analysis using various software packages, with data being pooled and reanalysed repeatedly until optimum stability and sense were achieved within each of the major vegetation groups.

	MG5a		MG5b		MG5c		MG5	
Festuca rubra	V	(1-8)	V	(2-8)	V	(2-7)	V	(1-8)
Cynosurus cristatus	V	(1-8)	V	(1-7)	V	(1-7)	V	(1-8)
Lotus corniculatus	V	(1-7)	V	(1-5)	V	(2-4)	V	(1-7)
Plantago lanceolata	V	(1-7)	V	(1-5)	IV	(1-4)	V	(1-7)
Holcus lanatus	IV	(1-6)	IV	(1-6)	V	(1-5)	IV	(1-6)
Dactylis glomerata	IV	(1-7)	IV	(1-6)	V	(1-6)	IV	(1-7)
Trifolium repens	IV	(1-9)	IV	(1-6)	V	(1-4)	IV	(1-9)
Centaurea nigra	IV	(1-5)	IV	(1-4)	V	(2-4)	IV	(1-5)
Agrostis capillaris	IV	(1-7)	IV	(1-7)	V	(3-8)	IV	(1-8)
Anthoxanthum odoratum	IV	(1-7)	IV	(1-8)	V	(1-4)	IV	(1-8)
Trifolium pratense	IV	(1-5)	IV	(1-4)	IV	(1-3)	IV	(1-5)
Lolium perenne	IV	(1-8)	III	(1-7)	I	(1-3)	III	(1-8)
Bellis perennis	III	(1-7)	II	(1-7)	I	(4)	II	(1-7)
Lathyrus pratensis	III	(1-5)	I	(1-3)	I	(1)	II	(1-5)
Leucanthemum vulgare	III	(1-3)	I	(1-3)	II	(1-3)	II	(1-3)
Festuca pratensis	II	(1-5)	I	(2-5)	I	(1)	I	(1-5)
Knautia arvensis	I	(4)					I	(4)
Juncus inflexus	I	(3-5)					I	(3-5)
Galium verum	I	(1-6)	V	(1-6)			II	(1-6)
Trisetum flavescens	II	(1-4)	IV	(1-6)	II	(1-3)	III	(1-6)
Achillea millefolium	III	(1-6)	V	(1-4)	III	(1-4)	III	(1-6)
Carex flacca	I	(1-4)	II	(1-4)	I	(1)	I	(1-4)
Sanguisorba minor	I	(4)	II	(3-5)			I	(3-5)
Koeleria macrantha	I	(1)	II	(1-6)			I	(1-6)
Agrostis stolonifera	I	(1-7)	II	(1-6)	I	(6)	I	(1-7)
Festuca ovina			II	(1-6)			I	(1-6)
Prunella vulgaris	III	(1-4)	III	(1-4)	IV	(1-3)	III	(1-4)
Leontodon autumnalis	II	(1-5)	II	(1-3)	IV	(1-4)	III	(1-5)
Luzula campestris	II	(1-4)	II	(1-6)	IV	(1-4)	III	(1-6)
Danthonia decumbens	I	(2-5)	I	(1-3)	V	(2-5)	I	(1-5)
Potentilla erecta	I	(1-4)	I	(3)	V	(1-4)	I	(1-4)
Succisa pratensis	I	(1-4)	I	(1-5)	V	(1-4)	I	(1-5)
Pimpinella saxifraga	I	(1-4)	I	(1-4)	III	(1-4)	I	(1-4)
Stachys betonica	I	(1-5)	I	(1-4)	III	(1-4)	I	(1-5)
Carex caryophyllea	I	(1-4)	I	(1-3)	II	(1-2)	I	(1-4)
Conopodium majus	I	(1-4)	I	(1-5)	II	(2-3)	I	(1-5)
Ranunculus acris	IV	(1-4)	II	(1-4)	IV	(2-4)	III	(1-4)
Rumex acetosa	III	(1-4)	III	(1-4)	III	(1-3)	III	(1-4)
Hypochaeris radicata	III	(1-5)	II	(2-4)	III	(1-4)	III	(1-5)
Ranunculus bulbosus	III	(1-7)	II	(1-5)	III	(1-2)	III	(1-7)
Taraxacum officinale agg.	III	(1-4)	III	(1-4)	III	(1-3)	III	(1-4)
Brachythecium rutabulum	II	(1-6)	III	(1-4)	II	(2)	III	(1-6)
Cerastium fontanum	III	(1-3)	II	(1-3)	II	(1-3)	II	(1-3)
Leontodon hispidus	II	(1-6)	III	(2-4)	III	(1-5)	II	(1-6)
Rhinanthus minor	II	(1-5)	II	(1-4)	II	(1-3)	II	(1-5)
Briza media	II	(1-6)	III	(1-4)	III	(2-3)	II	(1-6)
Heracleum sphondylium	II	(1-5)	II	(1-3)	III	(1-3)	II	(1-5)
Trifolium dubium	II	(1-8)	II	(1-5)	I	(2)	II	(1-8)
Primula veris	II	(1-4)	II	(2-4)	I	(2)	II	(1-4)
Arrhenatherum elatius	II	(1-6)	II	(1-7)	I	(3-4)	II	(1-7)
Cirsium arvense	II	(1-3)	II	(1-4)	I	(1)	II	(1-4)
Eurhynchium praelongum	II	(1-5)	II	(1-4)	I	(1-2)	II	(1-5)
Rhytidiadelphus squarrosus	II	(1-7)	II	(1-5)	III	(1-4)	II	(1-7)
Poa pratensis	II	(1-6)	II	(2-5)			II	(1-6)
Poa trivialis	II	(1-8)	I	(1-3)	I	(1-2)	II	(1-8)
Veronica chamaedrys	II	(1-4)	I	(1-4)	I	(1)	II	(1-4)
Alopecurus pratensis	I	(1-6)	I	(1-4)	I	(1)	I	(1-6)
Cardamine pratensis	I	(1-3)	I	(1)	I	(3)	I	(1-3)
Vicia cracca	I	(1-4)	I	(1-3)	I	(1-2)	I	(1-4)
Bromus hordeaceus hordeaceus	I	(1-6)	I	(2-3)	I	(3)	I	(1-6)
Phleum pratense pratense	I	(1-6)	I	(1-5)	I	(1)	I	(1-6)

44

	MG5a		MG5b		MG5c		MG5	
Juncus effusus	I	(2-3)	I	(3)	I	(1-2)	I	(1-3)
Phleum pratense bertolonii	I	(1-3)	I	(1-3)	I	(1)	I	(1-3)
Calliergon cuspidatum	I	(1-5)	I	(2-4)	I	(3)	I	(1-5)
Ranunculus repens	II	(1-7)	I	(2)	II	(1-4)	I	(1-7)
Pseudoscleropodium purum	I	(1-5)	I	(3-4)	II	(2)	I	(1-5)
Ophioglossum vulgatum	I	(1-5)	I	(1)			I	(1-5)
Silaum silaus	I	(1-5)	I	(1-3)			I	(1-5)
Agrimonia eupatoria	I	(1-5)	I	(1-3)			I	(1-5)
Avenula pubescens	I	(2-5)	I	(2-5)			I	(1-5)
Plantago media	I	(1-4)	I	(1-4)			I	(1-4)
Alchemilla glabra	I	(2)	I	(3)			I	(2-3)
Alchemilla filicaulis vestita	I	(1-3)	I	(3)			I	(1-3)
Alchemilla xanthochlora	I	(1-3)	I	(2)			I	(2-3)
Carex panicea	I	(1-4)	I	(2-4)			I	(1-4)
Colchicum autumnale	I	(3-4)	I	(1-3)			I	(1-4)
Crepis capillaris	I	(1-5)	I	(3)			I	(1-5)
Festuca arundinacea	I	(1-5)	I	(3-5)			I	(1-5)

Figure 12 Floristic table for NVC community MG5 *Cynosurus cristatus–Centaurea nigra* grassland. See text for details.

The simple term *community* was used to describe the basic units that emerged from these data analyses. It is roughly equivalent to the *Association* of Continental phytosociologists of the Braun-Blanquet school, and what defines each community as unique is the particular combination of frequency and abundance values for all the species found in the samples grouped within it. It is very important to realise that many species which are frequent in a given vegetation type have characteristically low covers and, conversely, that many infrequent species can have high abundance when they do occur. Some communities were so unvarying that no sub-division was necessary but many were further split into sub-communities. A very small number of especially bulky and complex communities have a third level of sub-division, into variants.

The quantitative species data for the NVC communities and their sub-divisions is summarised in a standardised format in floristic tables (e.g. Figure 12). Each floristic table includes such vascular plants, bryophytes and macrolichens as occur with a frequency of 5% or more in any one of the sub-communities (or, for vegetation types with no sub-communities, within the community as a whole). Early tests showed that records of species occurring below this level could be considered as largely insignificant for characterising vegetation types, but rejecting more frequent species meant that valuable floristic information was lost.

Every table has the frequency and abundance values for the species arranged in columns for the community as a whole and any sub-communities. Vascular plants are not separated from cryptogams although, on tables for woodland communities, trees and shrubs are listed separately to try to communicate some of the detailed structural complexity of the vegetation. More important is the way in which the species are arranged in blocks to indicate their pattern of occurrence through the community.

As an example of how such information is organised, the following sections describe the structure of the floristic table for MG5 *Cynosurus cristatus–Centaurea nigra* grassland.

7.6. Constant species and associates

The first group of species in a table, *Festuca rubra* to *Trifolium pratense* in this example (see Figure 12), is made up of the community constants – those species which have an overall frequency in classes IV or V. Generally speaking, such plants tend to maintain their high frequency in each of any sub-communities, though there may be some measure of variation in their representation from one to the next: here, for example, *Plantago lanceolata* is somewhat less common in the last sub-community than the first two. *Holcus lanatus* and a number of others show the reverse pattern. More often, there are considerable differences in the abundance of these most frequent species: many of the constants can have very high covers, while others are more consistently sparse. Also, quite commonly, plants which are not constant can be numbered among the dominants in a vegetation type.

The last group of species on a table, *Ranunculus acris* to *Festuca arundinacea* here (see Figure 12), comprises the general associates of the community, sometimes referred to as companions. These are plants which occur in the community as a whole with frequencies of class III or less, though sometimes they rise to constancy in one or other of the sub-communities, as with *R. acris* in this vegetation. Certain of the companions are consistently common overall, such as *Rumex acetosa*; some are more occasional throughout, as with *Rhinanthus minor*; some are always scarce, for example *Calliergon cuspidatum*. Others, however, are more unevenly represented, such as *R. acris*, *Heracleum sphondylium* or *Poa trivialis*, though they do not show any marked affiliation to any particular sub-community. Again, there can be much variation in the abundance of these associates: *Rumex acetosa*, for example, though quite frequent, is usually of low cover, while *Arrhenatherum elatius* and some of the bryophytes, though more occasional, can be patchily abundant. *Alchemilla xanthochlora* is both uncommon among the samples and sparse within them.

7.7. Preferential and differential species

The intervening blocks of species in a floristic table include those plants which are distinctly more frequent within one or more of the sub-communities than the others. Such species are referred to as preferential, or differential where their affiliation is more exclusive. For example, the group *Lolium perenne* to *Juncus inflexus* is particularly characteristic of the first sub-community of the *Cynosurus–Centaurea* grassland (see Figure 12), although some species, such as *Leucanthemum vulgare* and, even more so, *Lathyrus pratensis*, are more strongly preferential than others – *Lolium*, for example, continues to be frequent in the second sub-community. It is important to realise that even uncommon plants can be good preferentials, as with *Festuca pratensis* here: it is not often found in the *Cynosurus–Centaurea* grassland, but when it does occur, it is generally in this first sub-community.

The species group *Galium verum* to *Festuca ovina* (see Figure 12) helps to distinguish the second sub-community from the first, though again there is some variation in the strength of association: *Achillea millefolium*, for example, is less markedly diagnostic than *Trisetum flavescens* and, particularly, *G. verum*. There are also important negative features too: although some plants typical of the first and third sub-communities, such as *Lolium* and *Prunella vulgaris*, remain quite common here, the disappearance of others, such as *Lathyrus pratensis*, *Danthonia decumbens*, *Potentilla erecta* and *Succisa pratensis*, is strongly diagnostic. Similarly, with the third sub-community, there is that same mixture of positive and negative characteristics. Also, among all the groups of preferentials, there is the same variation in abundance as is found among the constants and companions. Thus, some plants which can be very marked preferentials are always of rather low cover, as with *Prunella*. Others such as *Agrostis stolonifera*, though diagnostic at low frequency, can be locally plentiful.

7.8. Character species and fidelity

In phytosociological terms, the character species of a vegetation type are those plants which show fidelity to it: that is, they are strongly preferential. Certainly, among the NVC vegetation types, it is possible to recognise species which are faithful to a community or range of communities and such plants are discussed in the community descriptions. Generally, though, in contrast to Continental schemes, such species have not been used to structure the floristic tables because their fidelity is often a very complex affair. More particularly, species which are widely recognised as faithful to particular types of vegetation elsewhere in Europe often lose this clear affiliation in the more Atlantic climate of the British Isles.

8 Identifying vegetation types

8.1. Comparing new samples with the NVC

Many users of the NVC wish to apply the scheme to identify vegetation which they encounter in the field. This involves making a comparison between such stands and the plant communities summarised in the floristic tables and described in the community accounts. With practice, it is possible for surveyors to make such comparisons without recording any data at all, just as experienced field botanists learn to recognise plant species on sight without recourse to a polythene bag and a flora.

Where data is being collected only for the purpose of identification, it is also possible to make some economies in recording (see Section 10.8). When starting to use the NVC, however, or when beginning survey in a new area or among unfamiliar vegetation, or when records are necessary to validate an identification on file and as a periodic check on the quality of survey, it is important to take standard NVC samples and make a more scrupulous comparison of some kind.

8.2. Keys to the NVC

The basis of identifying vegetation types using the NVC is to find which of the floristic tables in *British Plant Communities* gives the best fit for records collected from new stands. Ideally, a number of standard NVC samples should be collected from a number of stands of what looks like the same vegetation on a site or sites. It has become customary to use just five samples: this is only because it is easier to calculate frequency values from multiples of five, and five is the minimum viable number for this. In fact, the more samples that are available, the better. More important is to ensure, where possible, that these are not from the same stand: such intensive replication multiplies local oddities and can confuse identification by increasing the number of species of high constancy in a floristic table.

Comparing the floristic data summarised in tables is made easier by using keys. However, with something as complex as vegetation, no key is going to provide an infallible shortcut to identification, simply a rough guide to the most likely possibilities. The keys published in *British Plant Communities*

rely on floristic (and, to a lesser extent, physiognomic) data and they demand a thorough knowledge of the British vascular flora (and, in a few cases, of some bryophytes and lichens): it is vital to acknowledge that it is not really possible to identify or understand vegetation without being able to recognise plant species. The keys never make primary use of habitat factors, though these can provide valuable confirmation of a diagnosis.

Because the major distinctions between the vegetation types in the NVC are based on inter-stand frequency, most questions in the keys are concerned with how common or scarce species are in the data being interrogated. A floristic table prepared from new samples is thus the best basis for a comparison. The closer the approach comes to using single samples, the less reliable the process of comparison will be.

Most of the keys in *British Plant Communities* are dichotomous, and notes are provided at particularly difficult points and where confusing transitions between vegetation types are likely to be encountered. Other kinds of keys have been developed as the NVC has become widely applied to plant communities in particular regions, for example, or to restricted groups of vegetation types. The hierarchies and bull's-eye key to woodlands developed by Whitbread and Kirby (1992) are good examples of possible approaches. For grassland and montane communities summary descriptions and dendrogram keys have been published (Cooper 1997). Guides are also now available for mires and heaths (Elkington *et al.* 2001), woodland (Hall *et al.* 2004), and all upland communities (Averis *et al.* 2004). Any aid which helps with the understanding of variation among vegetation is to be welcomed.

What must be remembered is that keys alone, however subtle, are not enough to confirm identification. It is always necessary to make a closer scrutiny of the floristic tables of the selected type and of possible alternative choices and to read the detailed descriptions of the composition and structure of the vegetation in *British Plant Communities*.

8.3. MATCH and TABLEFIT

MATCH (Malloch 1998) and TABLEFIT (Hill 1996) are two computerised keys to the NVC plant communities. They work in similar ways, making statistical comparisons between one or more NVC samples and the floristic tables in *British Plant Communities*, using simple similarity coefficients. Trials indicate that both pieces of software can give valuable help and each has its own advantages. TABLEFIT appears to work the better of the two with single samples and can take quantitative data into account. MATCH can use frequency values from floristic tables of new samples but only uses qualitative data to make the statistical comparison: quantitative data are listed afterwards as a visual comparison between the input data and the NVC type.

MATCH also has the advantage that it lists ten top choices of most similar vegetation types, whilst TABLEFIT lists five. And for each of these choices, MATCH can list species that are under- or over-represented in the new data as compared with the NVC type: such lists can be very informative about the reasons why the newly surveyed vegetation is different from the described NVC community.

Such computerised keys are sometimes called 'expert systems' and are beguiling in their speed and statistical authority. However, they are no substitute for the experience of the ecologist and should *never* be used alone to provide identifications. Like written keys, they are simply a guide to negotiating a way around a complex classificatory landscape and to understanding variation that, in reality, is extremely complex.

8.4. Atypical vegetation and local peculiarities

For beginners, it is disconcerting to discover how often actual vegetation, newly encountered in the field, differs from the plant communities defined in the NVC floristic tables. It must be remembered, first of all, that these tables consist of data assembled from different places, often widely dispersed, and then summarised in a way that inevitably generalises much fine detail of difference. The vegetation types represented in the tables are real and can be recognised in the field but the definitions are necessarily an abstraction to some degree.

Local peculiarities will therefore always be found and these will manifest themselves as a poor fit or low correlation coefficient in a key. Such differences are especially likely where sampling has been concentrated in one stand or on a small site where local peculiarities will be replicated among the data. What is required is confidence to

be able to select the nearest described type and expertise to be able to interpret the differences from it in an ecologically informative fashion. In terms of local distinctiveness in the landscape, it should always be remembered that an unusual abundance of a particular species or the unexpected presence of a rarity can be a bonus.

8.5. Intermediate vegetation types

Variation within and between vegetation types is more or less continuous and any classification simply recognises centres of distinctive association between species and sharpens up the differences between the plant communities that are so characterised. The NVC should therefore be regarded as a set of pigeonholes providing a convenient summary of a very complex field of variation. Such a framework is invaluable for making inventories and maps of vegetation where discrete units and boundaries are needed, and this is one major reason why the NCC commissioned a classification of plant communities and not a gradient-based descriptive scheme. However, this means that stands of vegetation intermediate in composition and structure between two (or more) NVC plant communities are commonly encountered in the field. Using a key, it can therefore be hard to discriminate between alternative choices; with a computerised key, two possible vegetation types may have more or less identical similarity coefficients.

It is very important to realise that such difficulties in identification often have an underlying ecological explanation and can be very informative. For example, survey may have been carried out near one of the major climatic boundaries in Britain, say between the warmer and drier south and east and the cooler and wetter north and west: it is therefore not surprising that vegetation on or around this divide will show some features similar to one grassland or woodland, others characteristic of their counterparts further north. Or again, sampling may have been carried out on thinning ombrogenous peat between deep deposits supporting an active blanket bog and a humic podzol carrying wet heath: the vegetation might therefore be expected to be transitional in character between these types. Likewise, where grazing has recently been withdrawn, the vegetation may show some features of grassland communities, others characteristic of developing scrub.

In the NVC, many of the sub-communities described represent such transitions between vegetation types related to climates, soils or treatments which are intermediate or in flux. The community accounts also often detail zonations and successions that are commonly encountered in the field. However, to create multiples of units to reflect every grade of variation would defeat the purpose of a classification and it has not been possible to describe every transition and intermediate in the text of *British Plant Communities*. Understanding such subtleties is one of the challenges of using the NVC.

8.6. New vegetation types

The original aim of the NVC was to cover all natural, semi-natural and major artificial habitats in Great Britain (but not Northern Ireland), covering virtually all terrestrial plant communities, and those of brackish and fresh waters, except where no vascular plants were the dominants. In fact, as described in the published volumes of *British Plant Communities*, geographical and floristic coverage of the project was somewhat patchy and uneven. A total of about 35,000 relevés was available for the project and their distribution on the 10x10 km National Grid is shown in Figure 1. The geographical gaps are clearly visible, particularly in Scotland where the original project was almost entirely dependent on existing data or samples being collected contemporaneously but by other workers.

This figure also shows that the intensity of sampling within squares was very variable, with many 10x10 km squares having fewer than five samples while some have over a hundred. This variation is only partly related to the diversity of vegetation types sampled within an area. While every effort was made during the three seasons of fieldwork to ensure that the team of five surveyors covered as much ground as possible (Rodwell 1991a), the intensity of sampling reflects a measure of convenience of access. The particular interests of external contributors, whether in distinct vegetation types or certain areas, is also seen in the intensity of coverage.

Unevenness of floristic coverage and some of the more obvious gaps were referred to in the accounts of relevant plant communities and in the general introductions to the major vegetation types in *British Plant Communities*. However, since publication, use of the NVC and comparison with European phytosociological classification systems have revealed that there are types of British vegetation which have still to be described. As a result, the JNCC commissioned a review of the coverage of the NVC in 1998 (Rodwell *et al.* 2000). This review has produced information on the current coverage of the NVC; identified both the known and likely gaps in the plant community descriptions; and placed these new types into the phytosociological scheme of the NVC. In addition, the upland guide (Averis *et al.* 2004) describes some vegetation types not included in the NVC, whilst the coverage of the woodland section has been reviewed in Goldberg (2003).

The JNCC intends to establish a code that will define rules for the description of new variation in the NVC. The code will provide minimum standards for the description of new communities or sub-communities and a formal process for their validation and publication. An expert committee will be established and given authority to validate the descriptions of new types and ensure that the standards of the code are met.

9 Descriptions of plant communities

9.1. NVC community names and codes

In general, NVC communities described in *British Plant Communities* have been named using two or more of the most frequent and abundant constants, with any sub-communities named using distinctive preferentials. However, no new latinised terminology or complex hierarchical taxonomy was devised for the vegetation types and the detailed rules of the *International Code of Phytosociological Nomenclature* (Barkman *et al.* 1986) were not followed. Only where a phytosociological synonym has already attained a measure of common usage has this been adopted as an alternative name for a community, as with the *Centaureo–Cynosuretum* O'Sullivan 1965 or the *Pinguiculo–Caricetum* Jones 1975. Sometimes examples of a vegetation type can be found where the species used for naming it are absent, for example *Quercus–Betula–Dicranum* woodland which lacks oak (see Plate 7).

Every NVC community also had a letter and number code, where the letter(s) abbreviate the major vegetation type and the number indicates the position in the sequence of community descriptions. The letter codes for the major vegetation types and their arrangement in *British Plant Communities* are as follows:

Volume 1

W   Woodlands and scrub

Volume 2

M   Mires

H   Heaths

Volume 3

MG   Mesotrophic grasslands

CG   Calcicolous grasslands

U   Calcifugous grasslands and montane communities

Volume 4

A   Aquatic communities

S   Swamps and tall-herb fens

Volume 5

SD   Shingle, strandline and sand-dune communities

SM   Salt-marsh communities

MC   Maritime cliff communities

OV   Vegetation of open habitats

A full listing of the communities described in the five volumes of *British Plant Communities* is given in the Appendix. To find out which volume contains a particular vegetation type the reader should consult the indexes and general introductions to the volumes of *British Plant Communities*, or the Phytosociological Conspectus in Volume 5 (Rodwell 2000).

Any sub-communities characterised are denoted by lower-case letters after the community number: a, b, c etc. Thus, the *Cynosurus cristatus–Centaurea nigra* grassland is MG5; the *Galium verum* sub-community, the second to be distinguished, is MG5b.

9.2. Synonymy

The synonymy section of each community description lists those names applied to the vegetation type in previous surveys, together with the authors of the names and the date of ascription. The list of synonyms is arranged chronologically and includes references to important unpublished studies and to accounts of Irish and Continental associations where these are obviously very similar to the NVC types. It is important to realise that very many such synonyms are inexact. Sometimes the NVC community corresponds to just part of a previously described vegetation type, in which case the initials *p.p.* (for *pro parte*) follow the name. In other cases, the NVC vegetation type can be wholly subsumed within an older, more broadly defined unit. Despite this complexity, however, this section, together with that on the affinities of the vegetation, should help readers translate the NVC into terms with which they may have been long familiar. A special attempt has been made to indicate correspondence with particularly popular existing schemes and to make sense of venerable but ill-defined terms such as 'herb-rich meadow', 'oakwood' or 'general salt-marsh'. Each volume of *British Plant Communities* has an index of the synonyms of all the vegetation types included.

9.3. Constant species

The list of the constant species of the community includes those vascular plants, bryophytes and lichens which are of frequency classes IV and V overall.

9.4. Rare species

This list comprises any rare vascular plants, bryophytes and lichens which have been encountered in the particular vegetation type, or which are reliably known to occur in it. In this context, 'rare' means, for vascular plants, an A-rating in the *Atlas of the British Flora* (Perring and Walters 1962), where scarcity is measured by occurrence in numbers of vice-counties, or inclusion on lists compiled by the NCC of plants found in less than a hundred 10x10 km squares of the UK National Grid. For bryophytes, recorded presence in under 20 vice-counties has been used as a criterion (Corley and Hill 1981), with a necessarily more subjective estimate for lichens.

9.5. Physiognomy

The first substantial section of text in each community description is an account of the floristics and physiognomy of the vegetation, which attempts to communicate its essential character in a way which a tabulation of data could never do. Thus, the patterns of frequency and abundance of the different species which characterise the community are here filled out by details of the appearance and structure of the vegetation, variation in dominance and the growth form of the prominent groups of species, the physiognomic contribution of subordinate plants, and how all these components relate to one another. There is information, too, on important phenological changes that can affect the vegetation through the seasons and an indication of the structural and floristic implications of the progress of the life cycle of the dominants, any patterns of regeneration within the community or obvious signs of competitive interaction between plants.

Much of this material is based on observations made during the NVC sampling programme, but it has often been possible to incorporate insights from previous studies, sometimes as brief interpretative notes, in other cases as extended treatments of, say, the biology of particular species such as *Phragmites australis* or *Ammophila arenaria*, the phenology of winter annuals or the demography of turf perennials. Such information helps demonstrate the value of this kind of descriptive classification as a framework for integrating all manner of autecological studies (Pigott 1984).

9.6. Sub-communities

Some indication of the range of floristic and structural variation within each community is given in the discussion of general physiognomy, but where distinct sub-communities have been recognised these are each given a descriptive section of their own. The sub-community name is followed by any synonyms from previous studies, and by a text which concentrates on pointing up the particular features of composition and organisation which distinguish it from the other sub-communities.

9.7. Habitat

An opening paragraph in this section of each community description attempts to summarise the typical habitat conditions which favour the development and maintenance of the vegetation types, and the major environmental factors which control floristic and structural variation within it. This is followed by as much detail as was available at the time of the survey about the impact of particular climatic, edaphic and biotic variables on the character and distribution of the community and of environmental threats to it. With climate, for example, reference is very frequently made to the influence on the vegetation of the amount and disposition of rainfall through the year, the variation in temperature season by season, differences in cloud cover and sunshine, and how these factors interact in the maintenance of regimes of humidity, drought or frosts. Then there can be notes of effects attributable to the extent and duration of snow-lie or to the direction and strength of winds, especially where these are icy or salt-laden.

Commonly, too, there are interactions between climate and geology that are best perceived in terms of variations in soils. Here again, full weight has been given to the impact of the character of the landscape and its rocks and superficials, their lithology and the ways in which they weather and erode in the processes of pedogenesis. As far as possible, standardised terminology has been

51

employed in the description of soils, trying at least to distinguish the major profile types with which each community is associated, and to draw attention to the influence on its floristics and structure of processes like leaching and podzolisation, gleying and waterlogging, parching, freeze-thaw and solifluction, and inundation by fresh- or salt-waters.

With very many of the communities distinguished, it is combinations of climatic and edaphic factors that determine the general character and possible range of the vegetation, but biotic influences are also often clearly of importance, and there are very few instances where the impact of man cannot be seen in the present composition and distribution of the plant communities. Thus, there is frequent reference to the role which treatments such as grazing, mowing and burning have on the floristics and physiognomy of the vegetation, to the influence of manuring and other kinds of eutrophication, of draining and re-seeding for agriculture, of the cropping and planting of trees, of trampling, or other disturbance, and of various kinds of recreation.

The amount and quality of the environmental information for interpreting such effects has been very variable. The NVC sampling itself provided just a spare outline of the physical and edaphic conditions at each location, but it was also possible to draw on the substantial literature on the physiology and reproductive biology of individual species, on the taxonomy and demography of plants, on vegetation history and on farming and forestry techniques. Debts of this kind are always acknowledged in the text and the accounts should indicate the benefits of being able to locate experimental and historical studies on vegetation within the context of an understanding of plant communities (Pigott 1982).

9.8. Zonation and succession

Mention is often made in the discussion of the habitat of the ways in which stands of communities can show signs of variation in relation to spatial environmental differences, or the beginnings of a response to temporal changes in conditions. Fuller discussion of zonations to other vegetation types follows, with a detailed indication of how shifts in soil, microclimate or treatment affect the composition and structure of each community, and descriptions of the commonest patterns and particularly distinctive ecotones, mosaics and site types in which it and any sub-communities are found. It has also often been possible to give some fuller and more ordered

account of the ways in which vegetation types can change through time, with invasion of newly available ground, the progression of communities to maturity, and their regeneration and replacement. Some attempt has been made to identify climax vegetation types and major lines of succession, but it is vital to be wary of the temptation to extrapolate from spatial patterns to temporal sequences. The results of existing observational and experimental studies have been incorporated, where possible, including some of the classic accounts of patterns and processes among British vegetation. In addition, attention is drawn to the great advantages of a reliable scheme of classification as a basis for the monitoring and management of plant communities (Pigott 1977).

9.9. Distribution

Throughout the accounts, actual sites and regions are referred to wherever possible, many of them visited and sampled by the NVC survey team, some the location of previous studies, the results of which have been redescribed in the terms of the classification erected. The habitat section also provides some indications of how the overall ranges of the vegetation types are determined by environmental conditions. A separate paragraph on distribution summarises what is known of the ranges of the communities and sub-communities, then maps show the location, on the 10x10 km National Grid, of the samples that are available for each. Much ground, of course, has been thinly covered, and sometimes a dense clustering of samples can reflect intensive sampling rather than locally high frequency of a vegetation type. However, all the maps included are accurate in their general indication of distributions, and this exercise should encourage the production of a comprehensive atlas of British plant communities. The conservation agencies continue to update the distribution maps for many NVC types; for example, Horsfield et al. (1996) published updated distribution maps for upland NVC communities in Scotland, which have subsequently been extended to all upland Britain in Averis et al. (2004); whilst Hall (1997) and Hall et al. (2004) produced updated maps of the distribution of woodland NVC communities.

9.10. Affinities

The last section of each community description considers the floristic affinities of the vegetation types in the scheme, and expands on any particular problems of synonymy with previously described assemblages. Here, too, reference is made to the equivalent or most closely related association in Continental phytosociological classifications and an attempt is made to locate each community in an existing alliance. Where the fuller account of British vegetation that we have now been able to provide necessitates a revision of the perspective on European plant communities as a whole, some suggestions are made as to how this might be achieved. These wider European affinities were summarised in the Phytosociological Conspectus in Volume 5 of *British Plant Communities* (Rodwell 2000).

9.11. Indexes of species in *British Plant Communities*

In each of the published volumes of *British Plant Communities*, the classification and description of the vegetation types are supported by an index of all species recorded in the floristic tables. The species are listed alphabetically with the code numbers of the NVC communities in which they occur. Typographic differences indicate whether a species is constant in a community or sub-community or whether it occurs less commonly. To some extent, the species indexes can function as additional keys to the vegetation types. More substantially, they provide an introduction to the ecological relationships between individual species, whether rare or common, and the variety of plant communities in which they occur.

10 Carrying out an NVC survey

10.1. Defining what an NVC survey is

Even before the publication of *British Plant Communities*, the NVC had become widely acknowledged as a standard for the description of vegetation in this country. Now it is almost universally accepted not just as a classificatory framework but also as providing a methodology for vegetation survey, not only within the countryside and conservation agencies but also among agriculture and forestry organisations, NGOs, local authorities and corporate industry. Vegetation surveys are undertaken by these interests for a wide variety of reasons – for site description and inventory, for assessment of nature conservation value or environmental impacts, for vegetation management or as part of monitoring programmes.

The NVC has now found a place in all of these and can provide useful protocols and standards. However, it does have limitations and there is, in fact, no formal understanding of what 'an NVC survey' should include, even for the simplest kind of site description and inventory.

10.2. Site description and vegetation inventories

The most widespread application of the NVC is for listing and describing the vegetation types and vegetation patterns found on sites which are of recognised interest for nature conservation or under threat from some kind of development. Such survey usually involves obtaining an overview of variation on the site, delimiting what appear to be discrete stands of vegetation and identifying the different types of vegetation in relation to the published descriptions of *British Plant Communities*. Often, a site map is produced as a graphic display of the extent and distribution of the various vegetation types represented. Where more than one site is included in a survey, there can be comparisons between the patterns of occurrence of the different plant communities, or the development of a more extensive overview of a district or region, with tabular summaries showing the distribution and extent of the vegetation types represented across the localities.

10.3. Getting a general overview of a site

In encouraging some discipline in this kind of survey, the first thing to stress is the importance of familiarisation with an entire site and its vegetation pattern. Good quality map coverage of the site and its surrounds, at an appropriate scale, and air photographs can be extremely useful, even where vegetation maps are not envisaged. They will help the surveyor orientate within the wider landscape, devise an economic route for an initial walkabout and delineate the extent of any stands of vegetation that are of interest.

The importance of walking over the site cannot be overemphasised, even where time is at a premium. As a useful rule of thumb, one hour spent in thoughtful general observation on a site will be worth a day's detailed sampling. Such a walkabout should aim to give an overall perspective on a site and provide detail on the extent and scale of variation in the vegetation cover and on the occurrence of any repeating patterns. More broadly, it should lay the foundations for understanding the relationships between vegetation and habitat on the site and the wider landscape context of the area being surveyed.

10.4. Delimiting different vegetation types

Experienced surveyors cannot help but recognise familiar vegetation types even in preliminary examination of a site, and where such reliable skills are available, a walkabout can itself provide an inventory and sketch map of plant communities and their patterns: this is one of the benefits of accumulating experience of NVC survey. Commonly, however, and particularly where more detailed documentary support is required in a survey, this stage involves simply the recognition of the major vegetation boundaries on a site. Again, it is important to stress that such recognition need entail no decision about what the vegetation types are: what matters here is to mark out stands that are more or less homogeneous in composition and structure so they can be subsequently sampled and the vegetation identified.

Air photographs can again be very useful for locating such boundaries in the terrain, provided their meaning is ground-truthed, and sketch maps of the disposition of the stands may be essential as a framework for subsequent sampling. Of course, not all the vegetation of a site may be of interest: many site surveys are comprehensive but others will concentrate on particular kinds of vegetation. Even where this is the case, however, it is important to sustain a broad overview of a site, because the general disposition and context of the stands of interest may be vital for understanding the character of the vegetation.

A thorough familiarisation will also enable repeating patterns of vegetation types to be identified. Where economies of time have to be made, it may be sufficient to sample one complex of such types as a representative unit of a broader landscape. Characterising such complexes is itself of value in understanding how vegetation types are interrelated on a bigger scale.

10.5. Sampling vegetation types

Recording representative quadrats of the vegetation types of a site, using the same methodology as developed for the NVC project itself, has become a key element in survey. Generally, such samples have been collected so as to identify vegetation types. Providing such validating data allows other individuals or teams

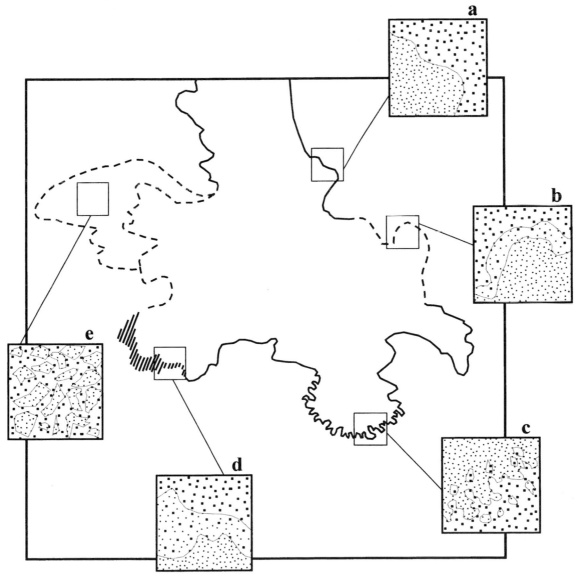

Figure 13 Representing complex vegetation boundaries on a map. The diagram shows a range of possible vegetation boundaries and ways to represent these: (a) a sharp boundary; (b) a diffuse boundary; (c) a convoluted boundary with islands; (d) a gradual transition; (e) a complex mosaic.

in an organisation or subsequent generations of workers to check the quality of survey and maintain standards of identification and interpretation. However, recording NVC samples can be time-consuming and it is very important to make a clear decision about the purpose of such recording and the value and fate of accumulated data before survey begins.

As far as the identification of vegetation types is concerned, this is easier, particularly for less skilled surveyors, if several rather than single NVC samples are collected from each type. The number five has no magical significance, but if this is the limit, such samples should be spread among several stands. Any multiplication and concentration of samples will replicate site or stand peculiarities in the vegetation.

Over and above such basic needs, a considered judgement should be made about the need to record further samples. Of course, there may be particular reasons for accumulating data – to record and understand fine variation in a diverse vegetation type, for example, to fill gaps where the NVC coverage is floristically or geographically patchy, or as part of some management or monitoring programme. Beyond such needs, there is no serious justification for accumulating NVC samples.

10.6. Closeness of fit to the NVC

A major preoccupation of surveys is to test how well the vegetation types sampled fit the published NVC classification and so identify the assemblages encountered on a site. An experienced surveyor will use his or her judgement to make such discriminations without needing to record any data. More usually, one or more samples of each vegetation type are checked against the described plant communities using written or computerised keys.

Any such identification should be able to characterise just how representative samples are of the defined type and to describe the nature and meaning of any distinctive features. It must be remembered that the floristic tables and descriptions in *British Plant Communities* provide broad generalisations from a national data set, so detailed sampling is certain to throw up variations ill-represented in the published scheme. In practice, most surveyors work at the sub-community level within the NVC: these seem to provide the most valuable scale of detail for site survey. However, even at this level, much interesting local diversity can be obscured by

simply quoting an 'answer' derived from a key. What is important is to use the NVC as a descriptive frame within which such information can be structured and understood. It is quite unacceptable simply to identify vegetation types with lists of highest correlation coefficients derived from statistical tests.

One other point to remember here is that, because the NVC has gained an authoritative place in the process of legal designation of sites, it is easy to regard the described vegetation types as normative for the purposes of protection, management and monitoring. A poor fit to such defined types is therefore seen as sub-standard. In fact, both well- and ill-fitting stands represent the real field of variation in a vegetation type which is summarised in a floristic table and description. In many cases, a 'poor' fit may indicate valuable local peculiarities, perhaps the very features for which a site was designated.

10.7. Vegetation mapping using the NVC

Many reports of NVC surveys now include maps of sites showing the disposition and extent of the vegetation types recognised. Again, however, there are no agreed standards for such maps, no graphic conventions nor common legend beyond the use of the code letters and numbers of the NVC communities or sub-communities. In practice, certain survey programmes, like those developed among the countryside and conservation agencies for woodlands, uplands and dune systems, have dominated and driven the progress of vegetation mapping using the NVC.

Most commonly, NVC vegetation maps have been produced at scales of 1:10,000 or 1:2,500. Both scales can provide valuable adjuncts to surveys but it should be remembered that the scale will determine the size of the minimum mappable unit. Even with a sharp hard pencil or narrow-gauge pen, it is only just possible to delineate an area of 1x1 mm on a map - this corresponds at a scale of 1:10,000 to a stand of 10x10 m. At such a scale, then, the accurate position and extent of, say, small flushes, springs, snow-beds, salt-pans, bog-pools or small ponds will be impossible to depict. Clearly, such fine-grained features are often of great interest but may have to be shown notionally and described in target notes.

Then there are often problems of depicting the character of boundaries, vegetation transitions or mosaics (see Figure 13), partly because of their size, but also because it is difficult to devise conventions subtle enough to portray gradual spatial shifts or complex patterns. Again, a notional compromise may have to be used on a map with detailed interpretation in accompanying text.

Accurate mapping of steeply-sloping ground, where complex mixtures of vegetation types can be disposed over ledges, crevices, rock faces and screes, is especially problematic because a great deal of variation may have to be summarised in negligible compass on a map. Then, the use of isodiametric sketches and profiles is especially valuable.

It is essential that each map has a comprehensive key attached to avoid any misinterpretation. Largely because of the cost and ease of production, most NVC vegetation maps have been made in black-and-white, sometimes with the use of various forms of shading. Colour mapping raises additional problems of co-ordinating with other standard sets of colour codes for particular vegetation types, as with those devised for Phase 1.

Since the publication of the NVC volumes, important advances have been made with the electronic capture of NVC maps and data, and the country nature conservation agencies have developed standards and specialised tools for this purpose. In addition, advances have been made in the use of remote-sensed images to map NVC communities (e.g. Meade 2004).

10.8. Acceptable economies in survey

Limited time and resources may make it necessary to economise within the general framework of NVC survey methodology. The first decision should be whether to collect any NVC samples at all: if vegetation types can be reliably identified without, then there may be little justification for such recording. Second, if samples are essential, then it may be sufficient to record qualitative (presence/absence) data for each sample, rather than Domin cover/abundance records. It should be remembered that most of the NVC communities and sub-communities are defined by inter-stand frequency, not by the abundance of the constituent species, so lists of species from samples may be enough to enable a preliminary identification to be made. Certainly, it is better in many cases to record several qualitative samples than one quantitative sample. However, for some vegetation types – such as swamps, salt-marsh and sand-dunes –

cover/abundance values can be essential for discriminating between the communities. Much more widely, quantitative information can be vital for understanding the dynamics, potential and trends in vegetation patterns.

A third possible economy is to omit recording species that are difficult to identify. Usually, bryophytes and lichens are the source of anxiety here. Certainly, in many cases, cryptogams are not essential for identifying vegetation types, even at sub-community level; and it is possible to characterise those situations where such identifications are essential for discriminating between assemblages. Again, though, such species can be informative about the general state of the vegetation and constitute an important element of diversity in their own right.

Beyond species records, precise details of the sample location, the date and author of the quadrat should never be omitted. Also, although environmental data, such as altitude, slope, aspect and geographical, soil and biotic information are not essential for identifying the vegetation type, this kind of detail may be invaluable for understanding the condition of the vegetation and the character of a site. Careful thought should be given as to whether relying only on simple quantitative values or summary terms for such parameters is sufficient. It is easy to forgo the challenge of making perceptive observations in the 'big box' of the NVC sample card, but these may comprise some of the most valuable accumulating experience of a surveyor or organisation.

10.9. Minimum data standards in NVC survey

In carrying out and reporting on an NVC survey, it is very important to be clear about how much of such standardised methodology has been employed and what economies, if any, are being made. Likewise, in drawing up a specification for survey or assessing the quality of work produced, such standards should be clearly understood, and explicitly detailed in contract briefs and assessments of reports. Maintaining minimum standards is a vital element in ensuring the quality of particular surveys, in accumulating skills and judgement in surveyors and teams and transferring the benefits of work and understanding to other groups and agencies or subsequent generations of staff.

Glossary of important terms

abundance (or cover/abundance):
the term used to describe how much of a species is present in a sample or stand, irrespective of how frequent or otherwise it is in moving from sample to sample or stand to stand.

associates:
species recorded in less than 60% of the samples of a vegetation type, in frequency class III (41–60%), II (21–40%) or I (1–20%).

association:
the term used in phytosociology to describe a plant community.

Braun-Blanquet scale:
a 5- or 6-point cover/abundance scale widely used in phytosociological survey elsewhere in Europe.

character species:
species with an especially marked fidelity for a particular vegetation type.

constants:
species recorded in over 60% of the samples of a vegetation type, either frequency class IV (61–80%) or V (81–100%).

differentials:
species with a more exclusive preference for a particular vegetation type, whether of low or high frequency.

Domin scale:
a ten-point scale of cover/abundance used to record the extent of species in NVC samples.

ecotone:
a transitional sequence of vegetation types often clearly related to an environmental gradient.

fidelity:
the faithfulness of a species to a particular vegetation type.

frequency:
the term used to describe how often a species is encountered in different samples or stands of vegetation, irrespective of its abundance.

homogeneous:
used to describe uniformity of floristic composition and structure in a stand of vegetation.

mosaic:
a pattern of two or more vegetation types disposed in intimate relationship to one another.

phytosociology:
the science of characterising and understanding plant associations through the collection and tabling of relevés.

preferentials:
species showing an affiliation to a particular vegetation type, whether of low or high frequency.

relevé:
the usual term (from French) describing an NVC-type sample of vegetation for phytosociological survey (in German, *Aufnahme*).

sample:
an NVC sample is a standardised record of the composition and structure of vegetation in a representative area of a homogeneous stand, together with basic site and environmental information.

stand:
an area of vegetation of any size.

synonym:
an existing name for part or all of a vegetation type.

References

Adam, P (1976) *Plant sociology and habitat factors in British saltmarshes.* Unpublished PhD thesis, University of Cambridge.

Averis, A M, Averis, A B G, Birks, H J B, Horsfield, D, Thompson, D B A and Yeo, M J M (2004) *An illustrated guide to British upland vegetation.* Joint Nature Conservation Committee, Peterborough.

Avery, B W (1980) *Soil classification for England and Wales (Higher Categories).* Soil Survey of England and Wales, Harpenden (Soil Survey Technical Monograph No. 14).

Barkman, J J, Moravec, J and Rauschert, S (1986) Code of phytosociological nomenclature, 2nd edn. *Vegetatio,* **67,** 145–195.

Birkinshaw, C (1991) *Derwent Ings ditch vegetation survey.* Nature Conservancy Council, Peterborough (England Field Unit Project Report, No. 47).

Birks, H J B (1969) *The late-Weichselian and present vegetation of the Isle of Skye.* Unpublished PhD thesis, University of Cambridge.

Birse, E L (1980) *Plant communities of Scotland: a preliminary phytocoenonia.* Macaulay Institute for Soil Research, Aberdeen.

Birse, E L (1984) *Phytocoenonia of Scotland: additions and revisions.* Macaulay Institute for Soil Research, Aberdeen.

Blockeel, T L and Long, D G (1998) *A checklist and census catalogue of British and Irish bryophytes.* British Bryological Society, Cardiff.

Bradshaw, M E and Jones, A V (1976) *Phytosociology in Upper Teesdale: guide to the vegetation maps of Widdybank Fell. With six accompanying maps.* Department of Extra Mural Studies, University of Durham, Durham.

Braun-Blanquet, J and Tüxen, R (1952) Irische pflanzengesellschaften. *Veröffentlichungen des geobotanischen institutes Rübel in Zürich,* **25,** 224–415.

Bridgewater, P (1970) *Phytosociology and community boundaries of the British heath formation.* Unpublished PhD thesis, University of Durham.

Bunce, R G H (1982) *A field key for classifying British woodland vegetation, Part I.* Institute of Terrestrial Ecology, Cambridge.

Charman, K (1981) *A botanical survey of ditches in selected areas of the North Kent Marshes.* Nature Conservancy Council, Peterborough (England Field Unit Project Report, No. 14).

Commission of the European Communities (1991) *CORINE biotopes.* Office for Official Publications of the European Communities, Luxembourg, for Commission of the European Communities.

Cooper, E A (1997) *Summary descriptions of National Vegetation Classification grassland and montane communities.* Joint Nature Conservation Committee, Peterborough (UK Nature Conservation, No. 14).

Cooper, E A, Crawford, I, Malloch, A J C and Rodwell, J S (1992) *Coastal vegetation survey of Northern Ireland.* Lancaster University, Unit of Vegetation Science Report to the Department of the Environment (Northern Ireland).

Corley, M F V and Hill, M O (1981) *Distribution of bryophytes in the British Isles.* British Bryological Society, Cardiff.

Dahl, E (1968) *Analytical key to British macrolichens.* 2nd edn. British Lichen Society, London.

Dahl, E and Hadač, E (1941) Strandgesellschaften der Insel Ostøy im Oslofjord. Eine pflanzensoziologische studie. *Nytt Magasin for Naturvidenskapene B,* **82,** 251–312.

Devilliers, P and Devilliers-Terschuren, J (1996) *A classification of Palaearctic habitats.* Council of Europe Publishing, Strasbourg (Environment and Nature Series, No. 78).

Doarks, C (1990) *Changes in the flora of grazing marsh dykes in Broadland, between 1972–74 and 1988–89.* Nature Conservancy Council, Peterborough (England Field Unit Project Report, No. 76.2).

Doarks, C and Leach, S (1990) *A classification of grazing marsh dyke vegetation in Broadland.* Nature Conservancy Council, Peterborough (England Field Unit Project Report, No. 76.5).

Doarks, C, Leach, S, Storer, J, Reid, S, Newlands, C and Kennison, G (1990) *An atlas of flora of grazing marsh dykes in Broadland.* Nature Conservancy Council, Peterborough (England Field Unit Project Report, No. 76.4).

Doarks, C and Storer, J (1990) *A botanical survey and evaluation of grazing marsh dyke systems in Broadland, 1988–89.* Nature Conservancy Council, Peterborough (England Field Unit Project Report, No. 76.3).

Elkington, T, Dayton, N, Jackson, D L and Strachan, I M (2001) *National Vegetation Classification: field guide to mires and heaths.* Joint Nature Conservation Committee, Peterborough.
http://www.jncc.gov.uk/page-2628

European Commission DG Environment (2003) *Interpretation manual of European Union habitats* (Version EUR 25). European Commission DG Environment, Brussels.

Géhu, J-M (1975) Aperçu sur les chênaies-hêtraies acidiphiles du Sud de l'Angleterre: L'exemple de la New Forest. In: *La végétation des forêts caducifoliées acidiphiles,* ed. by J-M Géhu), 133–140. Cramer, Leutershausen.

Glading, P (1980) *A botanical survey of ditches on the Pevensey Levels.* Nature Conservancy Council, Peterborough (England Field Unit Project Report, No. 25).

Goldberg, E (ed.) (2003) National Vegetation Classification – ten years' experience using the woodland section. *JNCC Report*, No. **335**. http://www.jncc.gov.uk/page-2348

Graham, G G (1971) *Phytosociological studies of relict woodlands in the north-east of England.* Unpublished MSc dissertation, University of Durham.

Hall, J E (1997) An analysis of National Vegetation Classification survey data. *JNCC Report*, No. **272**.

Hall, J E, Kirby, K J and Whitbread, A M (2004) *National Vegetation Classification: field guide to woodland.* Revised edn. Joint Nature Conservation Committee, Peterborough. http://www.jncc.gov.uk/page-2656

Haslam, S M (1978) *River plants.* Cambridge University Press, Cambridge.

Haslam, S M (1982) *Vegetation in British rivers.* Nature Conservancy Council, Peterborough.

Haslam, S M and Wolseley, P A (1981) *River vegetation: its identification, assessment and management.* Cambridge University Press, Cambridge.

Hennekens, S M and Schaminée J H J (2001) TURBOVEG, a comprehensive database management system for vegetation data. *Journal of Vegetation Science*, **12**, 589-591.

Hill, M O (1979) TWINSPAN – *a FORTRAN program for arranging multivariate data in an ordered two-way table by classification of the individuals and attributes.* Cornell University, New York.

Hill, M O (1996) TABLEFIT *version 1.0, for identification of vegetation types.* Institute of Terrestrial Ecology, Huntingdon.

Hodgson, J M (ed.) (1976) *Soil survey field handbook.* Soil Survey of England and Wales, Harpenden (Soil Survey Technical Monograph, No. 5).

Holmes, N T H (1983) *Typing British rivers according to their flora.* Nature Conservancy Council, Shrewsbury (Focus on Nature Conservation, No. 4).

Holmes, N T, Boon, P J and Rowell, T A (1999) *Vegetation communities of British rivers – a revised classification.* Joint Nature Conservation Committee, Peterborough.

Holmes, N T H, Lloyd, E J H, Potts, M and Whitton, B A (1972) Plants of the River Tyne and future water transfer scheme. *Vasculum*, **53**, 56–78.

Holmes, N T H and Whitton, B A (1977a) The macrophytic vegetation of the River Tees in 1975: observed and predicted changes. *Freshwater Biology*, **7**, 43–60.

Holmes, N T H and Whitton, B A (1977b) Macrophytes of the River Weaver 1966–76. *Naturalist*, **102**, 53–73.

Hooper, M D (1970) Dating hedges. *Area*, **4**, 63-65

Horsfield, D, Thompson, D B A and Tidswell, R (1996) *Revised atlas of upland NVC plant communities in Scotland.* Scottish Natural Heritage, Perth (Information and Advisory Note, No. 56).

Ivimey-Cook, R B and Proctor, M C F (1966) The plant communities of the Burren, Co. Clare. *Proceedings of the Royal Irish Academy B*, **64**, 211–301.

Jackson, D L and McLeod, C R (eds.) (2000) Handbook on the UK status of EC Habitats Directive interest features: provisional data on the UK distribution and extent of Annex I habitats and the UK distribution and population size of Annex II species. *JNCC Report*, No. **312**. http://www.jncc.gov.uk/page-2447

James, P W, Hawksworth, D L and Rose, F (1977) Lichen communities in the British Isles: a preliminary conspectus. *In: Lichen ecology*, ed. by M R D Seaward, 295–413. Academic Press, London.

Kirby, K J (1988) *A woodland survey handbook.* Nature Conservancy Council, Peterborough.

Klötzli, F (1970) Eichen-, Edellaub- und Bruchwälder der Britischen Inseln. *Schweizerischen Zeitschrift für Forstwesen*, **121**, 329–366.

Leach, S and Doarks, C (1991) *A botanical survey of ditches on coastal grazing marshes in Essex and Suffolk.* Nature Conservancy Council, Peterborough (England Field Unit Project Report, No. 49).

Leach, S, Doarks, C, Reid, S and Newlands, C (1991) *A botanical survey of ditches on Exminster Marshes SSSI, Devon.* Nature Conservancy Council, Peterborough (England Field Unit Project Report, No. 78).

Leach, S and Reid, S (1989) *A botanical survey of grazing marsh ditch systems on the North Norfolk coast.* Nature Conservancy Council, Peterborough (England Field Unit Project Report, No. 77).

MacDonald, A, Stevens, P, Armstrong, H, Immirzi, P and Reynolds, P (1998) *A guide to upland habitats: surveying land management impacts.* Scottish Natural Heritage, Battleby.

McVean, D N and Ratcliffe, D A (1962) *Plant communities of the Scottish Highlands. A study of Scottish mountain, moorland and forest vegetation.* HMSO, London (Monographs of the Nature Conservancy, No. 1).

Malloch, A J C (1970) *Analytical studies of cliff-top vegetation in south-west England.* Unpublished PhD thesis, University of Cambridge.

Malloch, A J C (1988) VESPAN II. University of Lancaster, Lancaster.

Malloch, A J C (1998) MATCH version 2. University of Lancaster, Lancaster.

Malloch, A J C (1999) VESPAN III. University of Lancaster, Lancaster.

Meade, R (ed.) (2004) *Proceedings of the Peterborough Remote Sensing Workshop 30 September 2004.* English Nature, Peterborough.

Mucina, L, Rodwell, J S, Schaminée, J H J and Dierschke, H (1993) European Vegetation Survey: current state of some national programmes. *Journal of Vegetation Science*, **4**, 429–483.

Nature Conservancy Council (1989) *Guidelines for selection of biological SSSIs.* Joint Nature Conservation Committee, Peterborough http://www.jncc.gov.uk/page-2303

Nature Conservancy Council, England Field Unit (2003) *Handbook for Phase 1 habitat survey – a technique for environmental audit.* Revised reprint. Joint Nature Conservation Committee, Peterborough.

Palmer, M (1989) *A botanical classification of standing waters in Great Britain.* Nature Conservancy Council, Peterborough.

Palmer, M (1992) *A botanical classification of standing waters in Great Britain.* 2ⁿᵈ edn. Joint Nature Conservation Committee, Peterborough.

Palmer, M A, Bell, S L and Butterfield, I (1992) A botanical classification of standing waters in Britain: applications for conservation and monitoring. *Aquatic Conservation: Marine and Freshwater Ecosystems*, **2**, 125–143.

Perring, F H and Walters, S M (eds.) (1962) *Atlas of the British flora.* Nelson, London and Edinburgh, for Botanical Society of the British Isles.

Peterken, G F (1981) *Woodland conservation and management.* Chapman & Hall, London.

Peterken, G F (1993) *Woodland conservation and management.* 2ⁿᵈ edition. Chapman & Hall, London.

Pigott, C D (1977) The scientific basis of practical conservation: aims and methods of conservation. *Proceedings of the Royal Society of London, Series B*, **197**, 59–68.

Pigott, C D (1982) The experimental study of vegetation. *New Phytologist*, **90**, 389–404.

Pigott, C D (1984) The flora and vegetation of Britain: ecology and conservation. *New Phytologist*, **98**, 119–128.

Poore, M E D (1955) The use of phytosociological methods in ecological investigations, III: Practical application. *Journal of Ecology*, **43**, 606–651.

Purvis, O W, Coppins, B J, Hawksworth, D L, James, P W and Moore, D M (1992) *The lichen flora of Great Britain and Ireland.* Natural History Museum Publications and British Lichen Society, London.

Purvis, O W, Coppins, B J and James, P W (1994) *Checklist of lichens of Great Britain and Ireland.* British Lichen Society, London.

Rackham, O (1980) *Ancient woodland: its history, vegetation and uses in England.* Arnold, London.

Rackham, O (2003) *Ancient woodland: its history, vegetation and uses in England.* New edition. Castlepoint Press, Dalbeattie.

Reid, S, Newlands, C and Leach, S (1989) *A new classification of Broadland dyke vegetation.* Nature Conservancy Council, Peterborough (England Field Unit Project Report, No. 76.1).

Rodwell, J S (ed.) (1991a) *British Plant Communities, Vol. 1: woodlands and scrub.* Cambridge University Press, Cambridge.

Rodwell, J S (ed.) (1991b) *British Plant Communities, Vol. 2: mires and heaths.* Cambridge University Press, Cambridge.

Rodwell, J S (ed.) (1992) *British Plant Communities, Vol. 3: grasslands and montane communities.* Cambridge University Press, Cambridge.

Rodwell, J S (ed.) (1995) *British Plant Communities, Vol. 4: aquatic communities, swamps and tall-herb fens.* Cambridge University Press, Cambridge.

Rodwell, J S (1997) The NVC and monitoring. *Countryside Council for Wales, Contract Science Report*, No. **200**.

Rodwell, J S (ed.) (2000) *British Plant Communities, Vol. 5: maritime communities and vegetation of open habitats.* Cambridge University Press, Cambridge.

Rodwell, J S, Dring, J C, Averis, A B G, Proctor, M C F, Malloch, A J C, Schaminée, J N J and Dargie, T C D (2000) Review of coverage of the National Vegetation Classification. *JNCC Report*, No. **302**. Joint Nature Conservation Committee, Peterborough. http://www.jncc.gov.uk/page-2312

Rodwell, J S, Pignatti, S, Mucina, L and Schaminée, J H J (1995) European Vegetation Survey: update on progress. *Journal of Vegetation Science*, **6**, 759–762.

Rodwell, J S, Schaminée, J H J, Mucina, L, Pignatti, S, Dring, J and Moss, D (2002) *The diversity of European vegetation: an overview of phytosociological alliances and their relationships to EUNIS habitats.* National Reference Centre for Agriculture, Nature and Fisheries, Wageningen (Rapport EC-LNV, nr. 2002/054).

Shimwell, D W (1968) *The phytosociology of calcareous grasslands in the British Isles.* Unpublished PhD thesis, University of Durham.

Spence, D H N (1964) The macrophytic vegetation of freshwater lochs, swamps and associated fens. *In: The vegetation of Scotland*, ed. by J H Burnett, 306–425. Oliver & Boyd, Edinburgh.

Stace, C (1997) *New flora of the British Isles*, 2ⁿᵈ edn. Cambridge University Press, Cambridge.

Stace, C (1999) *Field flora of the British Isles.* Cambridge University Press, Cambridge.

Tansley, A G (1939) *The British Islands and their vegetation.* Cambridge University Press, Cambridge.

Tutin, T G, Heywood, V H, Burges, N A, Valentine, D H, Walters, S M and Webb, D A (eds.) (1964) *Flora Europaea, Volume 1. Lycopodiaceae to Platanaceae.* Cambridge University Press, Cambridge.

Watling, R (1981) Relationships between macromycetes and the development of higher plant communities. *In: The fungal community: its organization and role in the ecosystem*, ed. by D T Wicklow and G C Carroll, 427–458. Marcel Dekker, New York and Basel.

Watling, R (1987) Larger Arctic-Alpine fungi in Scotland. *In: Arctic and Alpine mycology II*, ed. by G A Laursen, J F Ammirati and S A Redhead, 17–45. Plenum Publishing Corporation, New York.

Westhoff, V, Morzer Bruijns, M F and Segal, S (1959) The vegetation of Scottish pine woodlands and Dutch artificial coastal pine forests; with some remarks on the ecology of *Listera cordata. Acta Botanica Neerlandica*, **8**, 422-448.

Wheeler, B D (1975) *Phytosociological studies on rich fen systems in England and Wales.* Unpublished PhD thesis, University of Durham.

Whitbread, A M and Kirby, K J (1992) *Summary of National Vegetation Classification woodland descriptions.* Joint Nature Conservation Committee, Peterborough (UK Nature Conservation, No. 4).

Willems, J H (1978) Observations on north-west European limestone grassland communities: phytosociological and ecological notes on chalk grasslands of southern England. *Vegetatio*, **37**, 141–150.

Appendix:
Complete listing of
NVC communities and codes

From Volumes 1–5 of *British Plant Communities* (Rodwell 1991–2000)

Aquatic communities (Vol. 4)

A1	*Lemna gibba* community
A2	*Lemna minor* community
A3	*Spirodela polyrhiza–Hydrocharis morsus-ranae* community
A4	*Hydrocharis morsus-ranae–Stratiotes aloides* community
A5	*Ceratophyllum demersum* community
A6	*Ceratophyllum submersum* community
A7	*Nymphaea alba* community
A8	*Nuphar lutea* community
A9	*Potamogeton natans* community
A10	*Polygonum amphibium* community
A11	*Potamogeton pectinatus–Myriophyllum spicatum* community
A12	*Potamogeton pectinatus* community
A13	*Potamogeton perfoliatus–Myriophyllum alterniflorum* community
A14	*Myriophyllum alterniflorum* community
A15	*Elodea canadensis* community
A16	*Callitriche stagnalis* community
A17	*Ranunculus penicillatus* ssp. *pseudofluitans* community
A18	*Ranunculus fluitans* community
A19	*Ranunculus aquatilis* community
A20	*Ranunculus peltatus* community
A21	*Ranunculus baudotii* community
A22	*Littorella uniflora–Lobelia dortmanna* community
A23	*Isoetes lacustris/setacea* community
A24	*Juncus bulbosus* community

Calcicolous grasslands (Vol. 3)

CG1	*Festuca ovina–Carlina vulgaris* grassland
CG2	*Festuca ovina–Avenula pratensis* grassland
CG3	*Bromus erectus* grassland
CG4	*Brachypodium pinnatum* grassland
CG5	*Bromus erectus–Brachypodium pinnatum* grassland
CG6	*Avenula pubescens* grassland
CG7	*Festuca ovina–Hieracium pilosella–Thymus praecox/pulegioides* grassland
CG8	*Sesleria albicans–Scabiosa columbaria* grassland
CG9	*Sesleria albicans–Galium sterneri* grassland
CG10	*Festuca ovina–Agrostis capillaris–Thymus praecox* grassland
CG11	*Festuca ovina–Agrostis capillaris–Alchemilla alpina* grass-heath
CG12	*Festuca ovina–Alchemilla alpina–Silene acaulis* dwarf-herb community
CG13	*Dryas octopetala–Carex flacca* heath
CG14	*Dryas octopetala–Silene acaulis* ledge community

Heaths (Vol. 2)

H1	*Calluna vulgaris–Festuca ovina* heath
H2	*Calluna vulgaris–Ulex minor* heath
H3	*Ulex minor–Agrostis curtisii* heath
H4	*Ulex gallii–Agrostis curtisii* heath

H5	*Erica vagans–Schoenus nigricans* heath
H6	*Erica vagans–Ulex europaeus* heath
H7	*Calluna vulgaris–Scilla verna* heath
H8	*Calluna vulgaris–Ulex gallii* heath
H9	*Calluna vulgaris–Deschampsia flexuosa* heath
H10	*Calluna vulgaris–Erica cinerea* heath
H11	*Calluna vulgaris–Carex arenaria* heath
H12	*Calluna vulgaris–Vaccinium myrtillus* heath
H13	*Calluna vulgaris–Cladonia arbuscula* heath
H14	*Calluna vulgaris–Racomitrium lanuginosum* heath
H15	*Calluna vulgaris–Juniperus communis* ssp. *nana* heath
H16	*Calluna vulgaris–Arctostaphylos uva-ursi* heath
H17	*Calluna vulgaris–Arctostaphylos alpinus* heath
H18	*Vaccinium myrtillus–Deschampsia flexuosa* heath
H19	*Vaccinium myrtillus–Cladonia arbuscula* heath
H20	*Vaccinium myrtillus–Racomitrium lanuginosum* heath
H21	*Calluna vulgaris–Vaccinium myrtillus–Sphagnum capillifolium* heath
H22	*Vaccinium myrtillus–Rubus chamaemorus* heath

Mires (Vol. 2)

M1	*Sphagnum auriculatum* bog pool community
M2	*Sphagnum cuspidatum/recurvum* bog pool community
M3	*Eriophorum angustifolium* bog pool community
M4	*Carex rostrata–Sphagnum recurvum* mire
M5	*Carex rostrata–Sphagnum squarrosum* mire
M6	*Carex echinata–Sphagnum recurvum/auriculatum* mire
M7	*Carex curta–Sphagnum russowii* mire
M8	*Carex rostrata–Sphagnum warnstorfii* mire
M9	*Carex rostrata–Calliergon cuspidatum/giganteum* mire
M10	*Carex dioica–Pinguicula vulgaris* mire
M11	*Carex demissa–Saxifraga aizoides* mire
M12	*Carex saxatilis* mire
M13	*Schoenus nigricans–Juncus subnodulosus* mire
M14	*Schoenus nigricans–Narthecium ossifragum* mire
M15	*Scirpus cespitosus–Erica tetralix* wet heath
M16	*Erica tetralix–Sphagnum compactum* wet heath
M17	*Scirpus cespitosus–Eriophorum vaginatum* blanket mire
M18	*Erica tetralix–Sphagnum papillosum* raised and blanket mire
M19	*Calluna vulgaris–Eriophorum vaginatum* blanket mire
M20	*Eriophorum vaginatum* blanket and raised mire
M21	*Narthecium ossifragum–Sphagnum papillosum* valley mire
M22	*Juncus subnodulosus–Cirsium palustre* fen-meadow
M23	*Juncus effusus/acutiflorus–Galium palustre* rush-pasture
M24	*Molinia caerulea–Cirsium dissectum* fen-meadow
M25	*Molinia caerulea–Potentilla erecta* mire
M26	*Molinia caerulea–Crepis paludosa* mire
M27	*Filipendula ulmaria–Angelica sylvestris* mire
M28	*Iris pseudacorus–Filipendula ulmaria* mire
M29	*Hypericum elodes–Potamogeton polygonifolius* soakway
M30	Related vegetation of seasonally-inundated habitats
M31	*Anthelia julacea–Sphagnum auriculatum* spring
M32	*Philonotis fontana–Saxifraga stellaris* spring
M33	*Pohlia wahlenbergii* var. *glacialis* spring
M34	*Carex demissa–Koenigia islandica* flush
M35	*Ranunculus omiophyllus–Montia fontana* rill
M36	Lowland springs and streambanks of shaded situations
M37	*Cratoneuron commutatum–Festuca rubra* spring
M38	*Cratoneuron commutatum–Carex nigra* spring

Maritime cliff communities (Vol. 5)

MC1   *Crithmum maritimum–Spergularia rupicola* maritime rock-crevice community
MC2   *Armeria maritima–Ligusticum scoticum* maritime rock-crevice community
MC3   *Rhodiola rosea–Armeria maritima* maritime cliff-ledge community
MC4   *Brassica oleracea* maritime cliff-ledge community
MC5   *Armeria maritima–Cerastium diffusum* ssp. *diffusum* maritime therophyte community
MC6   *Atriplex prostrata–Beta vulgaris* ssp. *maritima* sea-bird cliff community
MC7   *Stellaria media–Rumex acetosa* sea-bird cliff community
MC8   *Festuca rubra–Armeria maritima* maritime grassland
MC9   *Festuca rubra–Holcus lanatus* maritime grassland
MC10   *Festuca rubra–Plantago* spp. maritime grassland
MC11   *Festuca rubra–Daucus carota* ssp. *gummifer* maritime grassland
MC12   *Festuca rubra–Hyacinthoides non-scripta* maritime bluebell community

Mesotrophic grasslands (Vol. 3)

MG1   *Arrhenatherum elatius* grassland
MG2   *Filipendula ulmaria–Arrhenatherum elatius* tall-herb grassland
MG3   *Anthoxanthum odoratum–Geranium sylvaticum* grassland
MG4   *Alopecurus pratensis–Sanguisorba officinalis* grassland
MG5   *Cynosurus cristatus–Centaurea nigra* grassland
MG6   *Lolium perenne–Cynosurus cristatus* grassland
MG7   *Lolium perenne* leys and related grasslands
MG8   *Cynosurus cristatus–Caltha palustris* grassland
MG9   *Holcus lanatus–Deschampsia cespitosa* grassland
MG10   *Holcus lanatus–Juncus effusus* rush-pasture
MG11   *Festuca rubra–Agrostis stolonifera–Potentilla anserina* grassland
MG12   *Festuca arundinacea* grassland
MG13   *Agrostis stolonifera–Alopecurus geniculatus* grassland

Vegetation of open habitats (Vol. 5)

OV1   *Viola arvensis–Aphanes microcarpa* community
OV2   *Briza minor–Silene gallica* community
OV3   *Papaver rhoeas–Viola arvensis* community
OV4   *Chrysanthemum segetum–Spergula arvensis* community
OV5   *Digitaria ischaemum–Erodium cicutarium* community
OV6   *Cerastium glomeratum–Fumaria muralis* ssp. *boraei* community
OV7   *Veronica persica–Veronica polita* community
OV8   *Veronica persica–Alopecurus myosuroides* community
OV9   *Matricaria perforata–Stellaria media* community
OV10   *Poa annua–Senecio vulgaris* community
OV11   *Poa annua–Stachys arvensis* community
OV12   *Poa annua–Myosotis arvensis* community
OV13   *Stellaria media–Capsella bursa-pastoris* community
OV14   *Urtica urens–Lamium amplexicaule* community
OV15   *Anagallis arvensis–Veronica persica* community
OV16   *Papaver rhoeas–Silene noctiflora* community
OV17   *Reseda lutea–Polygonum aviculare* community
OV18   *Polygonum aviculare–Chamomilla suaveolens* community
OV19   *Poa annua–Matricaria perforata* community
OV20   *Poa annua–Sagina procumbens* community
OV21   *Poa annua–Plantago major* community
OV22   *Poa annua–Taraxacum officinale* community
OV23   *Lolium perenne–Dactylis glomerata* community
OV24   *Urtica dioica–Galium aparine* community
OV25   *Urtica dioica–Cirsium arvense* community
OV26   *Epilobium hirsutum* community
OV27   *Epilobium angustifolium* community
OV28   *Agrostis stolonifera–Ranunculus repens* community
OV29   *Alopecurus geniculatus–Rorippa palustris* community

OV30	*Bidens tripartita–Polygonum amphibium* community
OV31	*Rorippa palustris–Filaginella uliginosa* community
OV32	*Myosotis scorpioides–Ranunculus sceleratus* community
OV33	*Polygonum lapathifolium–Poa annua* community
OV34	*Allium schoenoprasum–Plantago maritima* community
OV35	*Lythrum portula–Ranunculus flammula* community
OV36	*Lythrum hyssopifolia–Juncus bufonius* community
OV37	*Festuca ovina–Minuartia verna* community
OV38	*Gymnocarpium robertianum–Arrhenatherum elatius* community
OV39	*Asplenium trichomanes–Asplenium ruta-muraria* community
OV40	*Asplenium viride–Cystopteris fragilis* community
OV41	*Parietaria diffusa* community
OV42	*Cymbalaria muralis* community

Swamps and tall-herb fens (Vol. 4)

S1	*Carex elata* swamp
S2	*Cladium mariscus* swamp and sedge-beds
S3	*Carex paniculata* swamp
S4	*Phragmites australis* swamp and reed-beds
S5	*Glyceria maxima* swamp
S6	*Carex riparia* swamp
S7	*Carex acutiformis* swamp
S8	*Scirpus lacustris* ssp. *lacustris* swamp
S9	*Carex rostrata* swamp
S10	*Equisetum fluviatile* swamp
S11	*Carex vesicaria* swamp
S12	*Typha latifolia* swamp
S13	*Typha angustifolia* swamp
S14	*Sparganium erectum* swamp
S15	*Acorus calamus* swamp
S16	*Sagittaria sagittifolia* swamp
S17	*Carex pseudocyperus* swamp
S18	*Carex otrubae* swamp
S19	*Eleocharis palustris* swamp
S20	*Scirpus lacustris* ssp. *tabernaemontani* swamp
S21	*Scirpus maritimus* swamp
S22	*Glyceria fluitans* water-margin vegetation
S23	Other water-margin vegetation
S24	*Phragmites australis–Peucedanum palustris* tall-herb fen
S25	*Phragmites australis–Eupatorium cannabinum* tall-herb fen
S26	*Phragmites australis–Urtica dioica* tall-herb fen
S27	*Carex rostrata–Potentilla palustris* tall-herb fen
S28	*Phalaris arundinacea* tall-herb fen

Shingle, strandline and sand-dune communities (Vol. 5)

SD1	*Rumex crispus–Glaucium flavum* shingle community
SD2	*Honkenya peploides–Cakile maritima* strandline community
SD3	*Matricaria maritima–Galium aparine* strandline community
SD4	*Elymus farctus* ssp. *boreali-atlanticus* foredune community
SD5	*Leymus arenarius* mobile dune community
SD6	*Ammophila arenaria* mobile dune community
SD7	*Ammophila arenaria–Festuca rubra* semi-fixed dune community
SD8	*Festuca rubra–Galium verum* fixed dune grassland
SD9	*Ammophila arenaria–Arrhenatherum elatius* dune grassland
SD10	*Carex arenaria* dune community
SD11	*Carex arenaria–Cornicularia aculeata* dune community
SD12	*Carex arenaria–Festuca ovina–Agrostis capillaris* dune grassland
SD13	*Sagina nodosa–Bryum pseudotriquetrum* dune-slack community
SD14	*Salix repens–Campylium stellatum* dune-slack community

SD15	*Salix repens–Calliergon cuspidatum* dune-slack community
SD16	*Salix repens–Holcus lanatus* dune-slack community
SD17	*Potentilla anserina–Carex nigra* dune-slack community
SD18	*Hippophae rhamnoides* dune scrub
SD19	*Phleum arenarium–Arenaria serpyllifolia* dune annual community

Salt-marsh communities (Vol. 5)

SM1	*Zostera* communities
SM2	*Ruppia maritima* salt-marsh community
SM3	*Eleocharis parvula* salt-marsh community
SM4	*Spartina maritima* salt-marsh community
SM5	*Spartina alterniflora* salt-marsh community
SM6	*Spartina anglica* salt-marsh community
SM7	*Arthrocnemum perenne* stands
SM8	Annual *Salicornia* salt-marsh community
SM9	*Suaeda maritima* salt-marsh community
SM10	Transitional low-marsh vegetation with *Puccinellia maritima* annual *Salicornia* species and *Suaeda maritima*
SM11	*Aster tripolium* var. *discoideus* salt-marsh community
SM12	Rayed *Aster tripolium* stands
SM13	*Puccinellia maritima* salt-marsh community
SM14	*Halimione portulacoides* salt-marsh community
SM15	*Juncus maritimus–Triglochin maritima* salt-marsh community
SM16	*Festuca rubra* salt-marsh community
SM17	*Artemisia maritima* salt-marsh community
SM18	*Juncus maritimus* salt-marsh community
SM19	*Blysmus rufus* salt-marsh community
SM20	*Eleocharis uniglumis* salt-marsh community
SM21	*Suaeda vera–Limonium binervosum* salt-marsh community
SM22	*Halimione portulacoides–Frankenia laevis* salt-marsh community
SM23	*Spergularia marina–Puccinellia distans* salt-marsh community
SM24	*Elymus pycnanthus* salt-marsh community
SM25	*Suaeda vera* drift-line community
SM26	*Inula crithmoides* stands
SM27	*Ephemeral* salt-marsh vegetation with *Sagina maritima*
SM28	*Elymus repens* salt-marsh community

Calcifugous grasslands and montane communities (Vol. 3)

U1	*Festuca ovina–Agrostis capillaris–Rumex acetosella* grassland
U2	*Deschampsia flexuosa* grassland
U3	*Agrostis curtisii* grassland
U4	*Festuca ovina–Agrostis capillaris–Galium saxatile* grassland
U5	*Nardus stricta–Galium saxatile* grassland
U6	*Juncus squarrosus–Festuca ovina* grassland
U7	*Nardus stricta–Carex bigelowii* grass-heath
U8	*Carex bigelowii–Polytrichum alpinum* sedge-heath
U9	*Juncus trifidus–Racomitrium lanuginosum* rush-heath
U10	*Carex bigelowii–Racomitrium lanuginosum* moss-heath
U11	*Polytrichum sexangulare–Kiaeria starkei* snow-bed
U12	*Salix herbacea–Racomitrium heterostichum* snow-bed
U13	*Deschampsia cespitosa–Galium saxatile* grassland
U14	*Alchemilla alpina–Sibbaldia procumbens* dwarf-herb community
U15	*Saxifraga aizoides–Alchemilla glabra* banks
U16	*Luzula sylvatica–Vaccinium myrtillus* tall-herb community
U17	*Luzula sylvatica–Geum rivale* tall-herb community
U18	*Cryptogramma crispa–Athyrium distentifolium* snow-bed
U19	*Thelypteris limbosperma–Blechnum spicant* community
U20	*Pteridium aquilinum–Galium saxatile* community
U21	*Cryptogramma crispa–Deschampsia flexuosa* community

Woodlands and scrub (Vol. 1)

W1   *Salix cinerea–Galium palustre* woodland
W2   *Salix cinerea–Betula pubescens–Phragmites australis* woodland
W3   *Salix pentandra–Carex rostrata* woodland
W4   *Betula pubescens–Molinia caerulea* woodland
W5   *Alnus glutinosa–Carex paniculata* woodland
W6   *Alnus glutinosa–Urtica dioica* woodland
W7   *Alnus glutinosa–Fraxinus excelsior–Lysimachia nemorum* woodland
W8   *Fraxinus excelsior–Acer campestre–Mercurialis perennis* woodland
W9   *Fraxinus excelsior–Sorbus aucuparia–Mercurialis perennis* woodland
W10   *Quercus robur–Pteridium aquilinum–Rubus fruticosus* woodland
W11   *Quercus petraea–Betula pubescens–Oxalis acetosella* woodland
W12   *Fagus sylvatica–Mercurialis perennis* woodland
W13   *Taxus baccata* woodland
W14   *Fagus sylvatica–Rubus fruticosus* woodland
W15   *Fagus sylvatica–Deschampsia flexuosa* woodland
W16   *Quercus* spp.*–Betula* spp.*–Deschampsia flexuosa* woodland
W17   *Quercus petraea–Betula pubescens–Dicranum majus* woodland
W18   *Pinus sylvestris–Hylocomium splendens* woodland
W19   *Juniperus communis* ssp. *communis–Oxalis acetosella* woodland
W20   *Salix lapponum–Luzula sylvatica* scrub
W21   *Crataegus monogyna–Hedera helix* scrub
W22   *Prunus spinosa–Rubus fruticosus* scrub
W23   *Ulex europaeus–Rubus fruticosus* scrub
W24   *Rubus fruticosus–Holcus lanatus* underscrub
W25   *Ptcridium aquilinum–Rubus fruticosus* underscrub

Lightning Source UK Ltd.
Milton Keynes UK
UKOW021439290512

193535UK00001BA/10/P